Innovations in Process Engineering 2021

Held at the 2021 AIChE Annual Meeting

Boston, Massachusetts, USA and Online
7 - 11 November and 15 - 19 November 2021

ISBN: 978-1-7138-5686-3

Printed from e-media with permission by:

Curran Associates, Inc.
57 Morehouse Lane
Red Hook, NY 12571

Some format issues inherent in the e-media version may also appear in this print version.

Printed with permission by Curran Associates, Inc. (2022)

For permission requests, please contact AIChE
at the address below.

AIChE
120 Wall Street, FL 23
New York, NY 10005-4020

Phone: (800) 242-4363
Fax: (203) 775-5177

www.aiche.org

Additional copies of this publication are available from:

Curran Associates, Inc.
57 Morehouse Lane
Red Hook, NY 12571 USA
Phone: 845-758-0400
Fax: 845-758-2633
Email: curran@proceedings.com
Web: www.proceedings.com

Innovations in Process Engineering 2021

Held at the 2021 AIChE Annual Meeting

Boston, Massachusetts, USA and Online
7 - 11 November and 15 - 19 November 2021

AIChE
The Global Home of Chemical Engineers

TABLE OF CONTENTS

624308 Elucidating Metabolic and Regulatory Mechanisms of Microbial Aromatic Utilization for Lignin Valorization

Abstract:

Lignin represents a renewable resource whose bioconversion could displace petroleum-based processes [1, 2]. Research has been conducted to develop a hybrid platform to generate value-added bioproducts from lignin breakdown products (LBPs) obtained by thermo-catalytic depolymerization of waste lignin [1-3]. Traditional model organisms are not well suited for converting LBPs that consist of various toxic aromatic compounds, but non-model organisms have been identified as ideal candidates. One such host is *Rhodococcus opacus*, which has demonstrated high native tolerance to LBPs and the ability to improve its tolerance and consumption through adaptive evolution. However, the key challenges in such lignin upgrading include our limited understanding of microbial utilization of toxic LBPs at gene levels and limited tools to engineer this organism. To understand and maximize its metabolic potential, we have employed multi-omics approaches, providing a systems-level understanding of the complex metabolism of the wild-type and evolved strains [4-8]. Additionally, we have developed a genetic toolbox for *R. opacus* engineering [9-12]. Despite recent advances in our understanding of its versatile metabolism and available genetic tools, studies of its gene functions at gene levels are still lagging.

In this presentation, we will discuss our recent efforts to facilitate functional studies of this non-model organism using our genetic toolbox and engineer this promising chassis for lignin valorization. Specifically, we have identified nine putative transcription factor (TF) genes in proximity to gene clusters for degradation of aromatic compounds and confirmed the function of these transcription factors using a *gfp* reporter gene and multiple deletion analyses. Interestingly, cross-interactions between different aromatic compounds' consumptions have been observed at the gene regulation levels, implying complex gene regulatory and metabolic networks of this organism. Additionally, we have developed and optimized a gene repression system based on CRISPR interference (CRISPRi) [11], and we will discuss our CRISPRi tool's utility by demonstrating the inducible accumulation of muconate from aromatics [13]. Altogether, this work represents a considerable achievement to understand and engineer this promising chassis for lignin conversion into value-added chemicals.

[1] Davis, K., Moon, T. S., Tailoring microbes to upgrade lignin. *Current opinion in chemical biology* 2020, *59*, 23-29.

[2] Anthony, W. E., Carr, R. R., DeLorenzo, D. M., Campbell, T. P., *et al.*, Development of Rhodococcus opacus as a chassis for lignin valorization and bioproduction of high-value compounds. *Biotechnology for biofuels* 2019, *12*, 192.

[3] Chatterjee, A., DeLorenzo, D. M., Carr, R., Moon, T. S., Bioconversion of renewable feedstocks by Rhodococcus opacus. *Current opinion in biotechnology* 2020, *64*, 10-16.

[4] Henson, W. R., Campbell, T., DeLorenzo, D. M., Gao, Y., *et al.*, Multi-omic elucidation of aromatic catabolism in adaptively evolved Rhodococcus opacus. *Metabolic engineering* 2018, *49*, 69-83.

[5] Yoneda, A., Henson, W. R., Goldner, N. K., Park, K., *et al.*, Comparative transcriptomics elucidates adaptive phenol tolerance and utilization in lipid-accumulating Rhodococcus opacus

PD630. *Nucleic Acids Res.* 2016, *44*, 2240–2254.

[6] Roell, G. W., Carr, R. R., Campbell, T., Shang, Z., *et al.*, A concerted systems biology analysis of phenol metabolism in Rhodococcus opacus PD630. *Metabolic engineering* 2019, *55*, 120-130.

[7] Henson, W. R., Hsu, F.-F., Dantas, G., Moon, T. S., Foston, M., Lipid metabolism of phenol-tolerant Rhodococcus opacus strains for lignin bioconversion. *Biotechnology for biofuels* 2018, *11*, 339.

[8] Hollinshead, W. D., Henson, W. R., Abernathy, M., Moon, T. S., Tang, Y. J., Rapid metabolic analysis of Rhodococcus opacus PD630 via parallel C-metabolite fingerprinting. *Biotechnology and bioengineering* 2016, *113*, 91-100.

[9] DeLorenzo, D. M., Henson, W. R., Moon, T. S., Development of Chemical and Metabolite Sensors for Rhodococcus opacus PD630. *ACS Synthetic Biology* 2017, *6*, 1973–1978.

[10] DeLorenzo, D. M., Moon, T. S., Construction of Genetic Logic Gates Based on the T7 RNA Polymerase Expression System in Rhodococcus opacus PD630. *ACS Synthetic Biology* 2019, *8*, 1921-1930.

[11] DeLorenzo, D. M., Rottinghaus, A. G., Henson, W. R., Moon, T. S., Molecular Toolkit for Gene Expression Control and Genome Modification in Rhodococcus opacus PD630. *ACS Synthetic Biology* 2018, *7*, 727-738.

[12] DeLorenzo, D. M., Moon, T. S., Selection of stable reference genes for RT-qPCR in Rhodococcus opacus PD630. *Scientific reports* 2018, *8*, 6019.

[13] DeLorenzo, D.M., Diao, J., Carr, R., Hu, Y., Moon, T.S., An Improved CRISPR Interference Tool to Engineer Rhodococcus opacus. ACS Synthetic Biology. doi.org/10.1021/acssynbio.0c00591

628399 Depolymerization of Corn Stover Lignin with Bulk Molybdenum Carbide Catalysts

Abstract:

Depolymerization of lignin into aromatic compounds in high yield for production of fuels and chemicals is vital to realize an economically competitive biorefinery. In this study, undoped and nickel-doped bulk molybdenum carbides were synthesized and evaluated in reductive depolymerization of lignin model compound and corn stover lignin in the presence of hydrogen. The highest lignin monomer yield of 37.3% was obtained with Ni-Mo_2C and ethanol/water solvent. Nickel doping promoted the depolymerization performance of Mo_2C, by increasing the number of metallic sites while decreasing that of acidic sites. This change in active site distribution mitigated CC coupling reactions and coking. Ethanol addition to the water solvent also significantly suppressed CC coupling. Mixing carbides with zeolites decreased monomer yields, presumably due to excessive repolymerization of reactive intermediates over the acidic zeolite. This work highlights the feasibility of developing effective lignin depolymerization catalysts by fine-tuning the properties of molybdenum carbide catalysts and solvent systems.

631798 Developing a Multi-Target Alkali Sterilization Strategy to Facilitate Lignin Dispersion and Promote Biological Lignin Valorization

Abstract:

Value-added utilization of lignin holds great promise for realizing sustainable and economic biorefineries. Biological lignin valorization represents an emerging green approach to upgrade lignin under mild operation conditions without intensive energy input. However, due to its complex three-dimensional aromatic structure, lignin generally exhibits poor water solubility and inhomogeneous distribution in the aqueous fermentation medium, limiting its microbial conversion efficiency significantly. The aim of this study is to develop a multi-target alkali sterilization strategy to effectively enhance lignin dispersion and fermentation performance. Furthermore, the colloidal particle size and molecular structure variations of lignin during the sterilization were thoroughly investigated to reveal the mechanisms of enhanced bioconversion performance. Results showed that a complete aseptic effect was realized when mixing lignin medium at an initial pH 12.7 for 24 h, demonstrating the effectiveness of the alkali sterilization strategy. Dynamic light scattering analysis implied that the volume of colloidal lignin particles decreased by 96.3% by alkali sterilization compared to the conventional thermal sterilization. Moreover, alkali sterilization reduced lignin weight-average molecular weight by 23.0% and narrowed the molar-mass dispersity by 23.8%. The lignin substrates with more uniform distribution and lower molecular weight benefited microbial cell growth and metabolism synchronously. Compared to conventional thermally sterilized lignin, alkali sterilized lignin medium promoted *Rhodococcus opacus* PD630 cell growth, lignin degradation, and lipids production, which were increased by 309%, 30.3%, and 48.3%, respectively. These results clearly implied that alkali sterilization dramatically improved the lignin bioconversion performance. Overall, this study presents a facile and effective sterilization strategy to overcome inhomogeneous lignin distribution in aqueous fermentation mediums, showing great potentials as a platform technique to promote biological lignin valorization.

631969 Demethylated Lignin As the Interface Enhancer in FDM 3D Printing for PA12 Composites

Abstract:

Fused deposition modeling (FDM) is a widely used 3D printing technique. One of the big challenges of FDM is the weak interface bonding among the layers which limits the mechanical properties of the printed items. In the current work, lignin, the most abundant aromatic biopolymer, was modified by converting its methoxy group to hydroxyl group, forming more phenolic OH groups. Quantitative ^{31}P NMR along with ^{1}H NMR technique was used to track the changes of phenolic OH and methoxy group, and results showed that the content of phenolic -OH group increased from 2.86 to 4.76 mmol/g while the -OMe group dropped from 15.66 to 5.00 mmol/g. The modified lignin (M-Lignin) was then applied in the FDM 3D printing with polyamide 12 (PA12). Due to the enhanced interaction between PA12 and M-Lignin, the melt viscosity increased significantly for the PA12/M-lignin blend during the filament extrusion. For the mechanical performance of the printed samples, the PA12/M-lignin composites with 10 wt% of the M-lignin had the highest tensile strength around 52.7 MPa (46.2 MPa for pure PA) and the Young's modulus of 1.98 GPa (1.69 GPa for pure PA12), while the elongation at break remained 46% (58% for pure PA). In addition, the M-lignin could also be potential to prevent the aging of PA12 in practical use.

632781 Biorenewable Chemical Production from Pyrolysate By Lactic Acid Bacteria

Abstract:

Lignocellulosic biomass is an abundant source of carbon, especially in agricultural areas like Iowa. Fast pyrolysis is one method that can be used to create fermentable sugars from cellulosic sources, but product of this process, bio-oil, is rich in (1) levoglucosan, an anhydrosugar form of glucose, and (2) phenolic and other inhibitory compounds. Here we report the engineering of *Lactococcus lactis* to utilize purified levoglucosan for the production of enantiomerically pure L-lactic acid and, by expressing a heterologous three-gene pathway, 1,2-propanediol. We also report media conditions that allow the production of these chemicals from diluted bio-oil and strategies to engineer tolerance to bio-oil and other "dirty" substrates. Finally, we report the development of CRISPR-assisted genome-scale engineering and library creation tools in this important industrial organism.

637317 Roundtable Discussion for Public Input for the 2022-2026 National Strategic Plan for Advanced Manufacturing

Abstract:

You are invited to a roundtable discussion to provide public input into the development of the 2022-2026 National Strategic Plan for Advanced Manufacturing. The White House Office of Science and Technology Policy (OSTP) is required to develop quadrennial strategic plan for advanced manufacturing with stakeholder input from industry and academia under the America COMPETES Act. The first of these strategic plans was issued in 2018, and is available at the following link: https://www.manufacturing.gov/sites/default/files/2021-06/Advanced-Manufacturing-Strategic-Plan-2018.pdf. Members of the OSTP Advanced Manufacturing Subcommittee will be present during this discussion to receive public input.

637317 Roundtable Discussion for Public Input for the 2022-2026 National Strategic Plan for Advanced Manufacturing

Abstract:

You are invited to a roundtable discussion to provide public input into the development of the 2022-2026 National Strategic Plan for Advanced Manufacturing. The White House Office of Science and Technology Policy (OSTP) is required to develop quadrennial strategic plan for advanced manufacturing with stakeholder input from industry and academia under the America COMPETES Act. The first of these strategic plans was issued in 2018, and is available at the following link: https://www.manufacturing.gov/sites/default/files/2021-06/Advanced-Manufacturing-Strategic-Plan-2018.pdf. Members of the OSTP Advanced Manufacturing Subcommittee will be present during this discussion to receive public input.

625610 Modeling Adsorption of Organics-Laden Air from Wood Drying Using the Virtual Moving Bed Model

Abstract:

Air leaving a wood dryer contains organics which must be removed before the air can be released to the atmosphere. The current state-of-the-art in wood drying is the regenerative thermal oxidizer (RTO), which reduces the organics to carbon dioxide over a packed bed of ceramics (and sometimes iron catalyst) at high temperatures. [1,2] Sustainability and economic pressures suggest a more passive mode of capture, where the organics are condensed and recovered instead of burned. Captis Aire has developed a moving bed thermal adsorption process which can recover organics from air using a fluidized bed adsorber. However, models for such a process are lacking in commercial process simulators. Developed as a general shortcut model for cyclical adsorption processes, the virtual moving bed model previously developed can be also used to model the continuous flow unit and provide insights into its performance [3,4,5] using the concept of adsorption efficiency proposed by Smith and Westerberg. [6] Pure component adsorption of representative compounds were modeled with both the Langmuir-Freundlich and thermodynamic Langmuir (tL) isotherms, [7] which were extended to mixtures using their multicomponent counterparts: the empirical loading ratio correlation (LRC) [8] and real adsorbed solution theory (RAST-aNRTL). [9,10] These isotherm predictions are used in the virtual moving bed model using the one-efficiency model, with unity bed-state efficiencies for some components, to make predictions of the performance of the fluidized bed adsorber system.

Works Cited:

1. Milota, M. R., and Western Dry Kiln Association. (2000). Regulation and Control of Air Emissions.
2. Zhang, Z., Jiang, Z., and Shangguan, W. (2016). Low-temperature Catalysis for VOCs Removal in Technology and Application: a State-of-the-Art Review. *Catalysis Today*, *264*, 270 – 278.
3. Sees, M.D., Kirkes, T., and Chen, C.-C. (2020). A Simple and Practical Process Modeling Methodology for Pressure Swing Adsorption. *submitted.*
4. Sees, M.D. (2018, October). "Novel Steady-State Process Modeling Methodology for Pressure Swing Adsorption." Paper presented at the 2018 Annual Meeting of the American Institute of Chemical Engineers, Pittsburgh, PA.
5. Sees, M.D. (2020, November). "Modeling of Gas Separations by Pressure Swing Adsorption Using a Novel Steady-State Methodology." Paper presented at the 2020 Annual Meeting of the American Institute of Chemical Engineers, San Francisco [virtual], CA.
6. Smith, O.J. and A.W. Westerberg. *Eng. Sci.*, 1991, 48(12), 2967 – 2976.
7. Chang, C.K., Tun, H., and Chen, C.-C. (2020). An Activity-Based Formulation for Langmuir Adsorption Isotherm. *Adsorption*, 2020, 26(3), 375 – 386.
8. Yang, R. T. (1997). *Gas Separation by Adsorption Processes*(Vol. 1). World Scientific.
9. Kaur, H., Tun, H., Sees, M.D., and Chen, C.-C. (2019). Local Composition Activity Coefficient Model for Mixed-Gas Adsorption Equilibria. *Adsorption*, *25*(5), 951-964.
10. Tun, H., and Chen, C.-C. (2020). Prediction of Mixed-Gas Adsorption Equilibria from Pure Component Adsorption Isotherms. *AIChE Journal*, 1 – 9.

630902 Process Intensification of Ethylene Oxide Process Using Microfibrous Entrapped Catalyst

Abstract:

The ever-growing use of commodity products is causing great demand for high-volume raw material intermediates such as ethylene oxide, which are extensively used in plastic bottles, anti-freeze, sports gear, detergents, and paints. In 2019 alone, the United States produced 2.8 million metric tons of Ethylene Oxide (EO) and is projected to increase 3-4% per year over the next decade [1]. Traditionally, partial oxidation of ethylene via a silver catalyst is used for the production of ethylene oxide. It is one of the most energy-intensive and inefficient processes in the chemical process industry. Even after numerous studies aimed at increasing efficiency, the current state-of-the-art EO process produces a large amount of waste CO_2, usually 20%-25% of the ethylene raw material [2]. Even a small increase in efficiency through process intensification can significantly reduce the deleterious impact on the environment and improve economic performance by greater utilization of the raw materials.

The partial oxidation process of ethylene is highly exothermic, which requires a large number of small diameter tubes in the reactor to achieve efficient reactant flow distribution and fast heat removal. Moreover, localized hotspots can originate from flow channeling inside the reactor and lower selectivity, causing unwanted full oxidation of ethylene. Various studies have demonstrated that modified mass and heat transfer configurations inside the reactor can significantly change the downstream separation train, impacting the overall cost and environmental impact of the process. For example, Andreas et al. found distributed oxygen dosing along the reactor length is beneficial for the EO process than the conventional process [3]. Barecka et al. applied PI screening strategies using computer-aided molecular design (CAMD) tools to find suitable alternative separating agents after identifying the bottlenecked section of the EO process [4].

This work focuses on assessing the impact of employing Microfibrous Entrapped Catalyst (MFEC) structures inside the EO reactor on the downstream separation processes. The MFEC structure is a novel catalyst structure that consists of sintered micron-sized metal, glass, or polymer fibers with small catalyst particles entrapped inside [5, 6]. The MFEC structure would considerably enhance the intra-particle heat transfer [7] in the EO reactor, resulting in increased per pass conversion. A plant-scale simulation model of the conventional EO process has been developed using the Aspen Plus process simulator. The use of MFECs in the intensified reactor is modeled based on experimental data. Changes in reactant concentration inside the reactor caused higher per pass conversion of ethylene than the conventional EO process, and a model was developed for the new, more compact separation train. Next, a techno-economic analysis was performed to find the return on investment (ROI) of the intensified design compared with the base case simulation model. To identify and compare the processes' overall environmental impact, we also performed a life cycle impact assessment (LCA) with a contribution analysis of key impact factors using Life Cycle Assessment software (LCSoft).

References

[1] https://www.statista.com/statistics/974787/us-ethylene-oxide-production-volume/ visited on 24th March 2021.

[2] Gahanta M. (2012) Development of an economically viable H2O2-based, liquid-phase ethylene oxide technology: reactor engineering and catalyst development studies. Ph.D. Dissertation, University of Kansas, Lawrence, KS.

[3] Peschel, A., Jörke, A., Sundmacher, K., & Freund, H. (2012). Optimal reaction concept and plant wide optimization of the ethylene oxide process. Chemical engineering journal, 207, 656-674.

[4] Barecka, M. H., Skiborowski, M., & Górak, A. (2017). A novel approach for process retrofitting through process intensification: Ethylene oxide case study. Chemical Engineering Research and Design, 123, 295-316.

[5] Cheng, X., Yang, H., & Tatarchuk, B. J. (2016). Microfibrous entrapped hybrid iron-based catalysts for Fischer–Tropsch synthesis. Catalysis Today, 273, 62-71.

[6] Choudhury, H. A., Cheng, X., Afzal, S., Prakash, A. V., Tatarchuk, B. J., & Elbashir, N. O. (2020). Understanding the deactivation process of a microfibrous entrapped cobalt catalyst in supercritical fluid Fischer-Tropsch synthesis. Catalysis Today, 343, 112-124.

[7] Sheng, M., Gonzalez, C. F., Yantz Jr, W. R., Cahela, D. R., Yang, H., Harris, D. R., & Tatarchuk, B. J. (2013). Micro scale heat transfer comparison between packed beds and microfibrous entrapped catalysts. Engineering Applications of Computational Fluid Mechanics, 7(4), 471-485.

632131 A Process Intensification Synthesis Approach to Adsorption-Based Reactive Separation Systems

Abstract:

Adsorption-based reactive separation systems comprise an important category of process intensification applications which exploit the synergy between adsorption and reaction either using a single process unit (e.g., membrane reactor) or multi-column units operated in a periodic manner (e.g., simulated moving bed reactor) [1-2]. By integrating multiple tasks into a single modular unit (at different time scales), these technologies have been demonstrated to effectively shift the thermodynamic limitations to achieve high reaction conversions, reduce cost investment, and/or minimize process environmental impacts [3]. The computer-aided design and optimization of these processes rely mostly on unit-level rigorous models to directly solve large-scale partial differential algebraic equations (PDAE) or to fully discretize the PDAE spatial and temporal domains resulting in nonlinear programming problems [4-6]. Systematic synthesis approaches, which can efficiently incorporate the temporal domain and address the role of multifunctional materials in reactive adsorption, are still lacking.

In this work, we propose a systematic framework for the synthesis of intensified adsorption-based reactive separation systems based on the Generalized Modular Representation Framework (GMF) [7-8]. The reactive adsorption systems are represented as an aggregation of abstract phenomena-based mass- and/or heat- exchange building blocks leveraging a Gibbs free energy-based driving force constraints formulation to characterize the mass transfer feasibility via reaction, separation, and/or adsorption. The spatial distribution information in each modular building block, as well as the temporal dynamics, are extracted via Orthogonal/Radau Collocation on Finite Elements [9]. The proposed synthesis approach offers the advantages to: (i) synergize multi-functional phenomena without pre-postulation of equipment/flowsheet structures, (ii) find the optimal design parameters and operating strategies via a single mixed-integer nonlinear programming model, and (iii) provide a unified & computational efficient representation for evaluation of adsorption-based reactive separation systems. The applicability and versatility of the proposed framework will be showcased via two case studies: (i) a membrane reactor for water-shift reaction to demonstrate the spatial representation and efficient computational formulations using GMF, and (ii) a sorption-enhanced steam methane reforming reactor for hydrogen production to determine the optimal periodic operation strategy via superstructure optimization.

Reference

[1] Rodrigues, A. E., Wu, Y. J., Madeira, L. M., & Faria, R. (2017). Sorption enhanced reaction processes (Vol. 1). World Scientific.

[2] Tian, Y., & Pistikopoulos, E. N. (2019). Synthesis of operable process intensification systems: advances and challenges. Current Opinion in Chemical Engineering, 25, 101-107.

[3] Barelli, L., Bidini, G., Gallorini, F., & Servili, S. (2008). Hydrogen production through sorption-enhanced steam methane reforming and membrane technology: a review. Energy, 33(4), 554-570.

[4] Agarwal, A., Biegler, L. T., & Zitney, S. E. (2010). A superstructure-based optimal synthesis of PSA cycles for post-combustion CO2 capture. AIChE journal, 56(7), 1813-1828.

[5] Arora, A., Iyer, S. S., & Hasan, M. F. (2018). GRAMS: A general framework describing adsorption, reaction and sorption-enhanced reaction processes. Chemical Engineering Science, 192, 335-358.

[6] Tsay, C., Pattison, R. C., & Baldea, M. (2018). A pseudo-transient optimization framework for periodic processes: Pressure swing adsorption and simulated moving bed chromatography. AIChE Journal, 64(8), 2982-2996.

[7] Papalexandri, K. P., & Pistikopoulos, E. N. (1996). Generalized modular representation framework for process synthesis. AIChE Journal, 42(4), 1010-1032.

[8] Tian, Y., & Pistikopoulos, E. N. (2018). Synthesis of operable process intensification systems – Steady-state design with safety and operability considerations. Industrial & Engineering Chemistry Research, 58(15), 6049-6068.

[9] Algusane, T. Y., Proios, P., Georgiadis, M. C., & Pistikopoulos, E. N. (2006). A framework for the synthesis of reactive absorption columns. Chemical Engineering and Processing: Process Intensification, 45(4), 276-290.

624809 Development and Commissioning of a Modular and Integrated Apparatus for the Quasi-Continuous Production of Crystalline Particles

Abstract:

In almost every industrial sector, companies are faced with constantly changing challenges. In the chemical and pharmaceutical industry, these range from increasing global competition to existing but not exhausted resource and energy saving potentials, growing market volatility, short product life cycles as well as rising quality demands and the desire for individual products in small quantities [1–3]. To meet the challenges successfully and thus ensure the long-term success of the company, innovative solutions are required. In addition to a modular and small-scale design, these include converting batch wise operation into its continuous form, increasing the degree of automation and the integration of different process steps into a single process apparatus [4,5].

Taking these aspects into account, a novel plant concept for quasi-continuous particle synthesis and separation has been developed. A high degree of automation and the combination of different process steps (cooling crystallization, solid-liquid separation and contact drying) in one plant ensure significant energy and resource savings. Furthermore, an intelligent control system in combination with innovative measurement technology enables the targeted adjustment of particle properties and an increase in product quality and quantity.

The basis of the apparatus is a belt filter in which the vacuum trays below the filter medium are replaced by functional units. While there are temperature control units in the areas of crystallization and drying, filtration segments are installed in the solid-liquid separation section. The individual functional units can be arranged in any order, which in turn ensures a high degree of plant flexibility and makes it possible to adapt quickly to new material systems or changing customer requirements. The process chamber in which particle production takes place is located above the filter medium and passes through the process chain from left to right. Figure 1 shows the described apparatus concept schematically.

In addition to the development, the focus is on the commissioning of the plant. Thereby, the crystallization and filtration tests carried out with the model system sucrose prove the general functionality of the plant. The experiments demonstrate that the process parameters influence the properties of the produced particles. In this context, it is shown that the temperature profile during the cooling process influences both the particle size and the distribution width, and that larger crystals are formed as a result of a higher residence time in the crystallization zone. It is also found that the crystallization step slightly increases the resistance of the filter media. Compared to the overall resistance, however, this increase is marginal, which means that negative consequences for the overall process are not to be assumed.

In the course of the presentation, the basic apparatus concept and its technical implementation as well as the results of experimental considerations will be presented.

Acknowledgement: The authors would like to thank the German Federal Ministry for Economic Affairs and Energy for the financial support within the ENPRO initiative (support code: 03ET1652E).

References:

[1] Lier, S.; Wörsdörfer, D.; Grünewald, M. Wandlungsfähige Produktionskonzepte: Flexibel, Mobil, Dezentral, Modular, Beschleunigt. *Chemie Ingenieur Technik*, **2015**, 87, 1147–1158. 342

[2] Lier, S.; Paul, S.; Ferdinand, D.; Grünewald, M. Modulare Verfahrenstechnik: Apparateentwicklung für wandlungsfähige Produktionssysteme. *Chemie Ingenieur Technik*, **2016**, 88, 1444–1454.

[3] Bieringer, T.; Buchholz, S.; Kockmann, N. Future Production Concepts in the Chemical Industry: Modular - Small-Scale - Continuous. *Chem. Eng. Technol.*, **2013**, 36, 900–910.

[4] Fleischer-Trebes, C.; Krasberg, N.; Bramsiepe, C.; Kockmann, N. Planungsansatz für modulare Anlagen in der chemischen Industrie. *Chemie Ingenieur Technik*, **2017**, 89, 785–799.

[5] Löbnitz, L. Auslegung des Separationsprozesses und Entwicklung neuer Verfahrenskonzepte zur integrierten Produktion und Separation kristalliner Aminosäuren, **2020**.

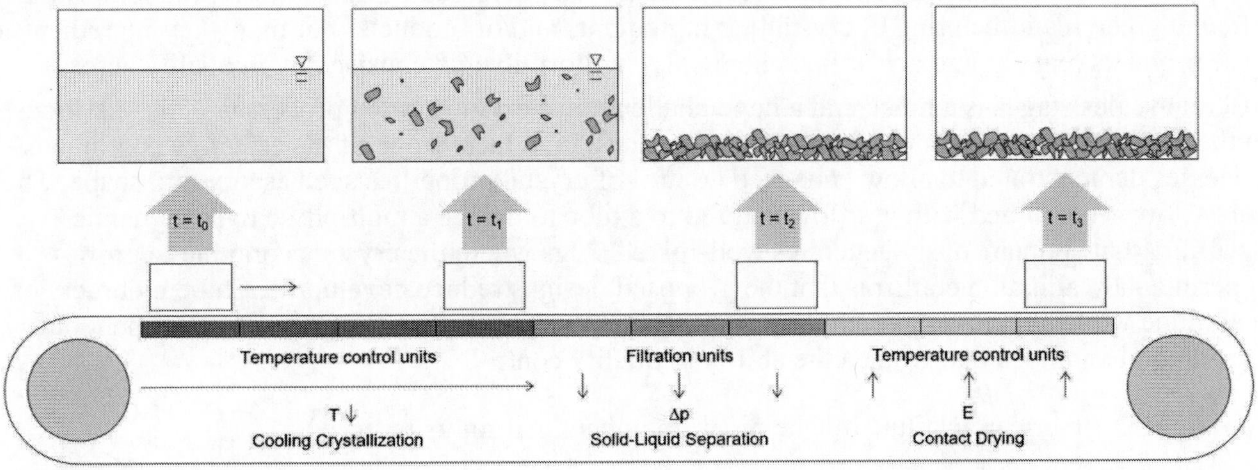

***Figure 1.** Apparatus concept and process procedure. At time $t = t_0$ there is an inoculated and saturated solution. By reducing the temperature T, a suspension has formed at $t = t_1$. The suspension is subsequently separated by an applied pressure difference Δp and leaves the solid-liquid separation zone at $t = t_2$. Due to the input of thermal energy E, only the dried product is left at $t = t_3$.*

623189 Scalable Modular Manufacturing of Crystals with Improved Uniformity

Abstract:

The pharmaceutical, material, and chemical industry has increased demands for reproducible and efficient green manufacturing of crystalline ingredients and/or products. For many key ingredients of these products (e.g., amino acids, fine chemicals), uniformity and consistency in quality across different scales/stages is a must, and a huge challenge for existing batch processes. This talk focuses on the slug-flow (or segmented flow) crystallizer design, which is one of the effective continuous processes demonstrated to allow enhanced control of crystal properties such as size and shape. The slurry flow is combined with gas flow and fed to a tube to induce a multiphase hydrodynamic instability that spontaneously generates well-mixed slugs where the crystals continue to grow. Experimental validation confirms that the proposed designs reduce development time by orders of magnitude while suppressing secondary nucleation, attrition, and clogging/fouling dominant but undesired phenomena that worsen the ability to quality control.

632968 Self-Heat Recuperative Heat Circulation System By Using Heat Pump Module

Abstract:

The self-heat recuperative thermal process consists of a material flow process with material stream composed of a series of elementary thermodynamic process modules (HR, HT, WR, WT) and a heat exchange module of energy stream (energy conversion module) with exergy destruction due to the heat transfer. Providing the compression work via work receiver (WR) to increase the temperature of the process fluid($T \rightarrow T+\Delta T$), the process heat Q can be circulated through the heat exchange module. The minimum work Wmin required for the circulation of process heat which compensate the exergy destruction is converted to the waste heat Qw of which anergy is equal to the exergy destruction T0Sgen as

In general, Wmin is much smaller than the process heat Q, leading to the drastic energy saving more than 50-90%. However, the direct compression by WR is unsuitable for the explosive gas. In such cases, the external heat pump module is preferable.

In this paper, a novel self-heat recuperative thermal process by using heat pump is described and its thermodynamic mechanism is elucidated in terms of exergy analysis such as a modularization method using the module expression flow, temperature-entropy, and energy conversion diagrams.

621487 Localized Heating of Microreactors for Alkane Dehydrogenation

Abstract:

Microwave heating offers a number of advantages, including spatially localized heating. By selectively absorbing the electromagnetic energy, a controlled temperature region can be created around a solid structure. In this talk, we discuss how to leverage this localized heating and demonstrate its effect on propane dehydrogenation in a microwave reactor. Using a 3D printed monolith as a heating scaffold, we model the spatially localized conversion of propane to propene by solving the electromagnetic, fluid flow/continuity, energy, and species equations with laboratory-based chemical kinetics. We will also present a reduced model using the Maxwell-Garnett formulation for the electromagnetic properties and a porous flow model. Using this simpler model, we elucidate the potential of localized heating to suppress the homogeneous side reactions (i.e., cracking) while promoting heterogeneous chemistries. Finally, we present experiments conducted using a microwave reactor that support model predictions, thereby highlighting a unique opportunity for reaction engineering that is absent in conventionally heated reactors.

624558 Physical Property Characterization of Ionic Liquids in HFC-32 and HFC-125 Using a High-Pressure View Cell

Abstract:

The current hydrofluorocarbon (HFC) refrigerants used in automobile air conditioning, home appliances and air-conditioners, and retail refrigerant systems are being phased out due to high global warming potentials, resulting in millions of kilograms of refrigerant needing to be recycled and repurposed. Many of these refrigerants are HFC mixtures, and if separated, the components may be repurposed or reused to prevent release of greenhouse gases into the atmosphere. Since most HFC mixtures are azeotropic, separation is difficult and energy intensive and thus no reasonable technologies exist today. Ionic liquids (ILs) offer a sustainable way to separate HFC refrigerants if an IL can be found with a large solubility difference between the components of an HFC mixture. This research is part of Project EARTH (Environmentally Applied Research Toward Hydrofluorocarbons) that is working to create an efficient and sustainable process for the separation of HFC refrigerants. The group has measured the solubility of HFC-125 (CHF_2CF_3) and HFC-32 (CH_2F_2), which make up the refrigerant R-410A (50-50 wt%), in selected ILs using a microbalance. The microbalance does not allow for visually observing what phase the IL might exist in when pressurized with the refrigerant, therefore this project utilizes a high-pressure view cell with quartz windows to visually study the solid-liquid phase change of imidazolium halide based ILs. In some cases, the ILs are solid at room temperature, but in the presence of HFC-125 and HFC-32 as a function of pressure melt and undergo a phase change. Understanding the phase behavior of an IL during HFC sorption allows for the management of heat in separation processes as heat of fusion can help reduce the heat of sorption. Other property measurements based on visual data, including melting point, the presence of metastable states, and occurrence of foaming, are taken in addition to phase change data. This presentation will be discussing how the physical properties of imidazolium halide based ILs change with the sorption of HFC-32 and HFC-125 to indicate the viability of the ILs for separation of the refrigerant R-410A.

629168 Material Selection and Design of Ionic Liquid-Based Extractive Distillation for Hydrofluorocarbons Separation

Abstract:

Hydrofluorocarbons (HFCs) and their mixtures are used as refrigerants for commercial and domestic cooling systems. Because of high global warming potential (GWP) of some of these mixtures, the Kyoto Protocol has recommended their gradual phase-out. There is an economic incentive to separate these HFC mixtures into their constituent components rather than venting or incinerating. However, many of these HFC mixtures are azeotropic in nature, and there is no commercial technology available to perform the separation. Extractive distillation, first proposed in the 1930s [1], is a technology that can separate close boiling azeotropic mixtures by using a higher boiling point entrainer. Ionic liquids (ILs) are attracting attention as entrainer for the extractive distillation of HFCs because of their negligible vapor pressure and chemical and thermal stability [2-3]. Due to significantly large number of possible ions combinations, theoretically, it is possible to have millions of ILs [4]. Selecting an appropriate IL for the separation of a particular HFC mixture is important for effective and economic distillation [5]. We have developed a multiscale process design, synthesis, and material selection framework called SPICE_ED for the synthesis of optimal extractive distillation process for separating HFC mixtures. SPICE_ED stands for Synthesis and Process Intensification of Chemical Enterprises involving Extractive Distillation. The framework is based on a building block-based novel representation of chemical phenomena, tasks, and operations in the chemical process industry [6-8]. From a candidate set, the framework can automatically select the most optimal IL that offers minimum energy requirement and/or has minimum processing cost. We generate process flowsheets and equipment designs considering the appropriate refrigerant and IL inlet trays, IL flow rates, reflux ratio, column operating conditions, and IL regeneration processes involving separate distillation, stripping, and flash operations. Finally, the framework can be utilized to map out the thermophysical properties of hypothetical ILs, thus directing the efforts towards discovering novel ILs with desired process economics. When searching for an optimal process for the separation of R-410A (50 wt% R-125 (C_2HF_5) and 50 wt% R-32 (CH_2F_2)), we obtained a heat-integrated optimized process that consumes 65% less energy and emits 72% less indirect CO_2 emission compared to previously reported designs with an existing IL [9]. In this presentation, we will also discuss how the framework can be employed for large-scale computational screening of ILs and other solvents for the separation of mixers with optimized process designs.

References:

[1] Randall, M. and Webb, W.A., 1939. Separation process. *Industrial & Engineering Chemistry, 31*(2), pp.227-230.

[2] Morais, A.R.C., Harders, A.N., Baca, K.R., Olsen, G.M., Befort, B.J., Dowling, A.W., Maginn, E.J. and Shiflett, M.B., 2020. Phase Equilibria, Diffusivities, and Equation of State Modeling of HFC-32 and HFC-125 in Imidazolium-Based Ionic Liquids for the Separation of R-410A. *Industrial & Engineering Chemistry Research, 59*(40), pp.18222-18235.

[3] Shiflett, M.B. and Maginn, E.J., 2017. The solubility of gases in ionic liquids. *AIChE Journal, 63*(11), pp. 4722-4737.

[4] Shiflett, M.B. and Yokozeki, A., 2006. Solubility and diffusivity of hydrofluorocarbons in

room-temperature ionic liquids. *AIChE Journal*, 52(3), pp.1205-1219.

[5] Lei, Z., Dai, C., Zhu, J. and Chen, B., 2014. Extractive distillation with ionic liquids: a review. *AIChE Journal*, *60*(9), pp.3312-3329.

[6] Demirel, S.E., Li, J. and Hasan, M.M.F., 2017. Systematic process intensification using building blocks. *Computers & Chemical Engineering*, *105*, pp.2-38.

[7] Demirel, S.E., Li, J. and Hasan, M.M.F., 2019. A General Framework for Process Synthesis, Integration, and Intensification. *Industrial & Engineering Chemistry Research*, *58*(15), pp.5950–5967.

[8] Demirel, S.E., Li, J., El-Halwagi, M.M. and Hasan, M.M.F., 2020. Sustainable Process Intensification using Building Blocks. *ACS Sustainable Chemistry & Engineering*, *8*(48), pp.17664–17679.

[9] Shiflett, M.B. and Yokozeki, A., 2006. Separation of difluoromethane and pentafluoroethane by extractive distillation using ionic liquid. *Chimica oggi*, *24*(2), pp.28-30.

631872 The Effect of Alkyl Chain Length on Imidazolium-Based Ionic Liquids and Choice of Halide Anions (Cl, Br, I) on Phase Equilibria for HFC-32, HFC-125, and Binary Mixtures (R-410A)

Abstract:

The current generation of refrigerants, hydrofluorocarbons (HFCs), are being phased out due to their high global warming potential (GWP) while the next generation of more environmentally friendly refrigerants, hydrofluoroolefins (HFOs), are developed and commercialized. Consequently, millions of tons of HFC mixtures will need to be recycled. Project EARTH (Environmentally Applied Research Toward Hydrofluorocarbons) is focused on developing efficient and selective separation processes for HFC mixtures. One promising approach is to use ionic liquids (ILs) as an entrainer to selectively separate azeotropic HFC mixtures like R-410A, which is composed of 50 wt.% HFC-125 (CHF_2CF_3) and 50 wt.% HFC-32 (CH_2F_2). The HFC-32 can be blended with HFOs to produce new products such as R-454B (68.9 wt.% HFC-32 and 31.1 wt.% HFO-1234yf ($CH_2{=}CFCF_3$)) and HFC-125 can be repurposed as a high-value fluorinated feedstock for production of new materials with a low-GWP.

This presentation will compare the effects of alkyl chain length on imidazolium-based ILs and anion substitution (Cl, Br, I) on the phase equilibria data for HFC-32 and HFC-125 and mixtures (e.g., R-410A). This presentation will also discuss ILs that are both solid and liquid at 25 °C. Phase equilibria data for HFC-32 and HFC-125 were measured using a Hiden Isochema IGA gravimetric balance and mixtures of HFC-32 and HFC-125 were measured using the first Hiden Isochema XEMIS Integral Mass Balance (IMB) method. The impact of alkyl chain length and anion on diffusion and selectivity will also be discussed for both pure HFC+IL and binary HFC+IL mixture measurements.

631738 Liquid-Liquid Equilibria of Ternary Systems of Thermally Robust Ionic Liquids and Hydrocarbon Mixtures

Abstract:

Hydrocarbon (HC) separations is an energy intensive process due to similar boiling points and formation of azeotropes. Here, we test different novel thermally robust perarylphosphonium and perarylsulfonium ionic liquids (ILs) for separating aromatic compounds from HC mixtures. Molecular dynamics (MD) simulations are used to provide a molecular-level understanding of these systems. Simulations of binary mixtures of toluene/ILs and *n*-heptane/ILs are performed to validate and study the miscibility behavior. Ternary mixtures consisting of IL/toluene/*n*-heptane at different concentrations of toluene in *n*-heptane are carried out to determine the amount of toluene absorbed into IL phase. Further, the fundamental interactions between aromatics and IL leading to this behavior are assessed through radial distribution functions and nearest neighbor analysis. Selectivity and distribution coefficients are determined to evaluate the affinity of toluene between IL and HC phase. The free energy of solvation ($\Delta G_{solvation}$) and transfer free energy ($\Delta G_{transfer}$) of toluene in different solvents are calculated using thermodynamic integration (TI) method. The results show that $\Delta G_{solvation}$ of toluene is more favorable in all the ILs compared to *n*-heptane. Overall the simulation results provide a molecular-level understanding of the solvent environment provided by these ILs in separating aromatics from hydrocarbon mixtures.

629325 The Deterministic Role of Hydrogen Bonding on Water Diffusivity in Methylimidazolium Halide Ionic Liquids and Ionogels

Abstract:

Polymeric ionogels, crosslinked gels swollen by ionic liquids (ILs), are useful vehicles for the release and storage of molecular solutes in next generation separation and delivery applications. Although rapid solute diffusion is often critical for performance, it remains challenging to predict diffusivities across multi-dimensional composition spaces. Recently, we showed that water (a model neutral solute and ubiquitous contaminant) diffuses through alkyl-methylimidazolium halide ILs by hopping between hydrogen bonding sites on relatively immobile cations. Here, we expand on this activated hopping mechanism in two significant ways. First, we demonstrate that water diffuses through poly(ethylene glycol) diacrylate ionogels via the same mechanism at a reduced rate. Second, we hypothesize that the activation energy barrier can be determined from relatively simple ^{1}H NMR chemical shift measurements of the proton responsible for H-bonding. This relationship enables water's diffusivity in ionogels of this class to be predicted quantitatively, requiring only (1) the composition-dependent diffusivity and Arrhenius behavior of a single IL, and (2) ^{1}H NMR spectra of the ionogels of interest. High-throughput microfluidic Fabry-Perot interferometry measurements verify prediction accuracy across a broad formulation space (4 ILs, $0 \leq x_{H2O} \leq 0.7$, $0 \leq \varphi_{PEGDA} \leq 0.66$). The predictive model may expedite IL material screening. Moreover, it intimates a powerful connection between solute mobility and hydrogen bonding, and suggests targets for rational design.

624169 Applications of Phytoremediation for Catalytic Enhancement of Biocarbon Towards Supercapacitors

Abstract:

The U.S. generates approximately two billion tons of mining waste annually, leading to inorganic heavy metal contaminates such as Ni, deposited into surrounding ecosystems at toxic levels. Phytoremediation offers a low-cost solution for environmental clean-up, particularly in impoverished areas, however current phytoremediation strategies generate unusable biomass due to the toxic metal contents. A novel solution for this unused waste lies in the area of energy storage, especially for supercapacitor applications. This work focuses on the cultivation of water hyacinths, a known metal hyperaccumulator, in various concentrations of Ni-doped water followed by synthesis of harvested biomass into high quality biocarbon. Biomass was subjected to thermal annealing at 500 °C for 1 h under Argon environment to cleave functional groups. Subsequent biochar was improved through thermochemical/catalytic activation at 800 °C for 1 h with KOH providing additional enhancement during biocarbon synthesis. Electrochemical results demonstrate a specific capacitance of 541 F g^{-1} making it suitable for use in the supercapacitor for electricity storage. Physical and electrochemical characterizations, including ICP, SEM, BET, ATR-FTIR, Raman, and XRD, cyclic voltammetry, chronopotentiometry, and EIS will be discussed. This work provides a value-added product and in-depth insights into naturally embedded catalyst, while simultaneously remediating contaminated mining sites.

624185 Electrochemical Studies on Graphitized Biocarbon Derived from Hydrothermally Liquefied Low Ash Content Corn Stover

Abstract:

Biomass-derived biocarbon has gained considerable attention due to its abundance, low-cost, and chemical structures conducive for energy storage applications, particularly supercapacitors. Lignocellulosic components embedded within the cell wall of biomass substrates can provide a fine template for enhanced ion storage, transport, and rate capabilities, desirable for electrochemical storage. Due to its abundant availability of agricultural residues such as Corn Stover, in this research we have utilized the homogenized low ash content corn stover milled and sieved to desired specifications. As obtained corn stover is converted to hydrochar via catalytic hydrothermal liquefaction in the presence of $Ni(NO_3)_2$ catalyst at 275°C and its morphological properties such as surface area, pore volume and diameter were enhanced in the presence of $ZnCl_2$, followed by thermal activation at 450°C. Later as derived hydrochar is pH neutralized and acid washed with dilute HCl to extract metal contaminants. Biochar underwent thermal carbonization at 850 °C for 3 h to enhance pore structures and graphitization, conducive for electrochemical applications. Electrochemical results demonstrated specific capacitances in excess of 300 F g^{-1}, comparable to literature values for corn stover-derived biocarbons. Physical and electrochemical characterizations, such as BET, ATR-FTIR, Raman, XRD, cyclic voltammetry and chronopotentiometry, will be discussed to give fundamental insights into biocarbon enhancement. The results obtained during the electrochemical characterization and studies will be presented.

625756 Electrocatalyst Development for Making Ammonia from Air and Water at Ambient Condition

Abstract:

Renewable power generated from wave energy has faced technological and cost barriers that have hindered its penetration into utility-scale electricity markets. As an alternative, the production of chemical fuels—for example, ammonia (NH_3), which has a high energy density (11.5 MJ/L) and facile storage properties—may open wave energy to new markets including ocean exploration and transportation. We developed a unique catalyst and tested its performance toward nitrogen reduction reaction using the rotating round disk electrode (RRDE) method. The current density increased 0.25 mA/cm^2 after switching the gas from Ar to N_2 and the production rate of NH_3 and Faradaic efficiency (FE) are in the top list of reported results.

627567 Ammonium Formate As a Safe, Energy-Dense Electrochemical Fuel Ionic Liquid

Abstract:

Hydrogen has been investigated widely as an energy carrier for clean electrons, however its low energy density and large energy losses incurred in compression have limited its use. For these reasons, solid and liquid carriers of hydrogen fuel have been investigated, including liquid ammonia, formic acid/formate salts, and methanol, to name a few. While these proposed fuels have significant potential, many of them do not meet the energy efficiency requirements necessary to outperform compressed hydrogen and enable facile transport and storage of renewable energy. For example, ammonia is an appealing energy carrier because of its high hydrogen content: 110 kg H_2/m^3 for liquid ammonia compared to 40 kg H_2/m^3 for compressed hydrogen at 700 bar and 288 K. It is already produced, transported, and stored at scale, and there has been recent progress in reducing the carbon footprint of ammonia production technologies. Formic acid is also appealing due to its hydrogen content (53 kg H_2/m^3), its ease of transport and storage since it is a liquid, and its potential to be synthesized efficiently from carbon dioxide, water, and renewable electricity. However, both ammonia and formic acid suffer from safety issues because they are caustic and corrosive, as well as difficulties with efficient release of energy at scale.

In this work, we propose ammonium formate, a combination of ammonia and formic acid, as an energy carrier. Not only is its production as cheap and simple as its constituent molecules, ammonium formate is a benign solid at ambient conditions, making it safe to transport and store. We demonstrate an electrochemical cell that takes advantage of the fact that ammonium formate forms an intrinsically conductive ionic liquid at 116°C; as an ionic liquid, ammonium formate has negligible vapor pressure and is significantly safer than either ammonia or formic acid alone. In this system, hydrogen is evolved at the cathode with near 100% Faradaic efficiency and formate is oxidized to carbon dioxide at the anode with near 100% Faradaic efficiency. In theory, ammonia could be oxidized to nitrogen in the same cell, however we observe that the operating temperature makes it so that ammonia gas evaporates from the cell and can be oxidized using a second, modular ammonia fuel cell. Because molten ammonium formate is conductive, the electrolyte and feed can approach 100% ammonium formate fuel, resulting in a large thermodynamic driving force. Overall, this system represents a new class of electrochemical fuel ionic liquids and results in modular release of H_2 with the potential for zero net carbon emissions.

632405 Enhanced CO_2 Electroreduction to CH_4 and C_2H_4 Via Selective Proton Transfer

Abstract:

CO_2-derived fuels, synthesized using renewable energy, present an attractive route towards fuel formation from sustainable energy. The role of proton transfer in dictating the efficiency and selectivity of the two-electron, two-proton reduction of CO_2 to CO has been well-studied. Indeed, impressive progress has been achieved in the production of CO from CO_2. However, the role of proton transfer to surface-bound CO intermediates in governing the selective electrosynthesis of higher order products beyond CO, such as methane or ethylene, remains poorly understood.

Here, we investigate the electrochemical reduction of CO at Cu electrodes in nonaqueous electrolytes at low temperatures. The choice of the electrolyte and reaction conditions allows fine-tuned control of the proton donor and the CO binding strength, enabling for the first time activation-controlled kinetic studies over an extended parameter range. Our studies lead to the surprising finding that increasing the concentration of CO serves to suppress methane production, in addition to hydrogen. In contrast, ethylene production remains unaffected by CO concentration and depends predominantly on the electrochemical potential.

These observations provide unprecedented insight into the mechanism of electrochemical CO and CO_2 reduction. Among others, we are able to show that the rate of methane and hydrogen formation is governed by the competition of CO and H for surface sites and that ethylene formation is taking place from sites which are saturated with CO, indicating that simply increasing the supply of CO may not be a viable strategy for improving CO and CO_2 reduction systems. Based on our results, we subsequently demonstrate how knowledge of the reaction mechanism can be exploited to selectively suppress hydrogen evolution relative to CO reduction by rational tuning of the proton donor.

635600 Biodirected Assembly of Ionomer Thin Films for Fuel Cells and Electrolyzers

Abstract:

Increasing evidence suggests that control of ionomer thin film structures on metal surfaces is pivotal for efficient performance in electrochemical devices such as fuel cells and electrolyzers. Unfortunately, the assembly of these thin film structures is difficult to control using conventional methods and ideal arrangements remain unknown. Engineered polypeptides have emerged as powerful biomolecular tools in electrode assembly because binding sites and polypeptide structures can be easily modulated by changing the amino acid sequence. However, no studies have been conducted showing polypeptides can be engineered to interact with ionomers, attach them to metal surfaces, and control their arrangement. Our lab has recently developed this technology, using an elastin-like polypeptide to bind to metals, and bind to acidic and basic ionomer via ionic interactions. We use a quartz crystal microbalance with dissipation to provide detailed information about the loading, thickness, and binding behavior of the polypeptide and ionomer layers. We also use atomic force microscopy and grazing-incidence small-angle X-ray scattering to understand the impact of the biomolecules on ionomer phase separation. Finally, we have begun to analyze the performance (ionic conductivity) of the assembled films using interdigitated electrodes and electrochemical impedance spectroscopy. Through these techniques, we show that 1) our biomolecular system is highly flexible, easily adapting to different materials, 2) the polypeptide sequence can dictate ionomer phase separated structures, and 3) this system can be used to improve the performance of ionomer thin films and gain structure-function understanding. Generally, our results demonstrate that engineered polypeptides are promising tools for ionomer control in electrode engineering.

635530 Parametric Testing, Analysis and Techno-Economic Assessment of 2-Eempa, a Single-Component Water-Lean Post-Combustion Capture Solvent

Abstract:

We present here, the design, comprehensive testing and technoeconomic assessment of N-(2-ethoxyethyl)-3-morpholinopropan-1-amine (2-EEMPA), a single-component water-lean solvent developed at PNNL. This presentation details the physical and thermodynamic properties of the solvent, including vapor–liquid equilibria, kinetics and viscosity measurements in addition to over 40 hours of continuous CO_2 capture performance in a laboratory scale system on simulated flue gas comprised of expected levels of H_2O, O_2, NOx and SOx. A full techno economic analysis of the capture process at industrial scale with corresponding projections of critical metrics was performed using Aspen Plus. Here, we present that 2-EEMPA can be optimized for efficiency, with a projected regeneration heat rate of 2.0 GJ per tonne CO_2 for post-combustion capture of coal-derived flue gas. Conversely, we detail how 2-EEMPA can be optimized for cost using a 2-stage flash regeneration, where projected total cost of capture of $37.6/tonne (in 2018 USD) CO_2 using the US Department of Energy's revised cost and performance assumptions. We close with a discussion on further cost reduction potential and preparation for upcoming slip-stream testing of this solvent at a 0.5 MW scale.

635542 Diversification of Basic Immobilized Amine Sorbents (BIAS) for Carbon Dioxide Capture

Abstract:

Carbon sequestration entails a multi-step process in which anthropogenic CO_2 emissions are captured from CO_2-laden process gas streams and permanently stored. Carbon capture is a critical step in the process and accounts for a considerable portion of the overall cost. Across the literature, a multitude of functionalized silica sorbents demonstrate high-capacity CO_2 capture and offer advantages over the benchmark liquid monoethanolamine (MEA) process, including reduced heat duty and improved CO_2 mass transfer. However, low sorbent stability still remains an issue. Basic immobilized amine sorbents (BIAS) developed at NETL were successfully tested at bench and pilot-scales, and demonstrated their promising potential for large scale CO_2 removal from post-combustion sources and ambient air. This is owed to the BIAS' optimized tradeoff between CO_2 capture capacity and thermal stability observed over multiple and even 100 adsorption-desorption cycles. Within the last few years, key technological milestones diversifying the commercialization potential of the BIAS were achieved. These include in house (bench scale, 10 g) and commercial (3 kg) pelletization of particles into rods and spheres for moving bed rectors; incorporation into a non-aqueous particle/silicone oil slurry regenerated by microwave irradiation for replacing/retro-fitting the MEA process; and spinning into hollow chemisorption fiber sorbents (CHEFS) for rapid temperature swing adsorption systems. The proven scalability of BIAS combined with the diverse portfolio of material architecture makes this technology ideal for industrial CO_2 capture applications, including for coal combustion and gasification power generating systems; landfill gas cleanup, enhanced oil recovery (EOR); natural gas sweetening, and life support systems.

636022 CO_2 Capture Modeling, Energy Savings, and Heat Pump Integration

Abstract:

While amine-based CO_2 capture systems have been designed since 1927, concern over anthropomorphic climate change sparked renewed interest in the 1980's. Subsequent research has led to significant advances such as well-fit thermodynamic models, rigorous rate-based distillation simulations, and improved process designs. The aims of this presentation are to (1) summarize proposed process improvements; (2) demonstrate the strong and complex interactions of energy-saving schemes; (3) report an optimal process design integrating all of these improvements in Aspen Plus, and then evolutionally simplify the design through simulation results. In particular, we propose an optimum energy-saving design (U.S. Patent No. 10,751,664) that uses an absorption-driven heat pump together with a distributed cross heat exchanger and a stripper vapor condensate rerouting in order to reduce both the cooling and heating utility consumptions. The resulting predicted solvent regeneration energy for the integrated system is 1.67 GJ/tonne CO2 captured. China Petroleum and Chemical Corporation (SINOPEC) is currently implementing the process to recover 1 million tons per year of CO_2 from the power plant flue gas in the Shengli Oil Field, Shangdong Province, for tertiary oil recovery.

624825 Process Intensification for Direct Conversion of Biomass-Based Syngas to High Octane Gasoline

Abstract:

The conversion of lignocellulosic-derived carbon sources via the creation of a CO_2 rich syngas stream provides means to produce liquid hydrocarbon fuels, with advancements intended to create a cost-competitive method of sequential conversion via methanol and dimethyl ether intermediates. Previously developed methods with Cu/Beta zeolites enabled the conversion of dimethyl ether to linear and branched hydrocarbons, with high selectivities towards products with high-octane fuel properties.[1] Through process intensification, use of a single reactor to convert syngas to methanol and DME, then further to hydrocarbons, allows for reduced capitol and operation cost for the same chemistries that typically use 3 reactors instead of one. Utilization of commercially available methanol synthesis catalyst (Megamax 800, Clariant, CZA) and methanol dehydration catalyst (g-Al_2O_3) with the Cu/Beta catalyst allowed for conversion into hydrocarbon products in this single reactor set up. Reactor configuration with CZA and alumina catalysts mixed with or stacked in a bed physically above the Cu/Beta catalyst provided a system for the direct conversion of the syngas to hydrocarbons with improved yields and selectivities with higher net C_1 conversions for stacked bed configurations. Through control of the process conditions, greater conversion of C1 oxygenate intermediates (i.e., methanol and dimethyl ether) was achieved with concomitant increase in selectivities towards gasoline and jet fuel range olefin and paraffin products. Lower hydrocarbon number products and less naphthenes were observed for the conversion of syngas compared to DME conversion on only the Cu/Beta system with prominent selectivities observed for C_4, C_5, and C_7 hydrocarbons.[2] The DME and methanol formation rates, with understanding of the equilibria for both processes, were determined as important factors to allow for improved performance of the Cu/Beta. Determination of the factors which allow for tuning of selectivities and yields created an intensified process which allows for a "market-responsive" biorefinery design, which can produce high octane gasoline or jet fuel range hydrocarbons to meet demands for a more sustainable route to liquid fuels.

References:

(1)Schaidle, J. A.; Ruddy, D. A.; Habas, S. E.; Pan, M.; Zhang, G.; Miller, J. T.; Hensley, J. E. Conversion of Dimethyl Ether to 2,2,3-Trimethylbutane over a Cu/BEA Catalyst: Role of Cu Sites in Hydrogen Incorporation. *ACS Catal.* **2015**, *5* (3). 1794–1803.

(2)Ruddy, D. A.; Hensley, J. E.; Nash, C. P.; Tan, E. C. D.; Christensen, E.; Farberow, C. A.; Baddour, F. G.; Van Allsburg, K. M.; Schaidle, J. A. Methanol to High-Octane Gasoline within a Market-Responsive Biorefinery Concept Enabled by Catalysis. *Nat. Catal.* **2019**, *2* (7), 632–640.

630926 Carbon Oxide-Thermic Oxidation Process for Synthesis of Porous Silicon-Carbon Composite Anodes for Lithium-Ion Batteries

Abstract:

Carbon Oxide-Thermic Oxidation Process for Synthesis of Porous Silicon-Carbon Composite Anodes for Lithium-ion Batteries

Mohamed A. Elharati[a], Younghwan Cha[b], Qusay Bkour[a], Steven R. Saunders[a], M. Grant Norton[b], Min-Kyu Song[b,*], Su Ha[a,*]

[a]Voiland School of Chemical Engineering and Bioengineering, Washington State University, Pullman, WA 99164-2710, USA

[b]School of Mechanical and Materials Engineering, Washington State University, P.O. Box 642920, Pullman, WA 99164-2920, USA

Abstract

The silicon-based compounds have been investigated as a promising anode material for lithium-ion (Li-ion) batteries. In this work, we show that a novel carbon oxide thermic oxidation (CO_x-OP) process is not only a facile and low-cost technique to synthesize porous silicon-carbon (Si-C) composites for high-energy Li-ion batteries but also is the novel utilization technology of greenhouse gas. Furthermore, we demonstrate that the morphology/microstructure and performance of Si-C composites can be controlled by using different gas compositions (CO, CO_2, and H_2).

In-situ high-temperature X-ray diffraction analyses were performed to elucidate the phase transition during CO_x-OP. The pulsing experiments were conducted to illustrate the mechanism of the CO_x-OP process. The electron microscopy and X-ray microanalysis were carried out to investigate the effects of processing conditions on the morphology and microstructures of Si-C. The TGA and Raman analysis were performed to determine the amount and type of carbonaceous materials.

The synthesized porous Si-C via CO_x-OP process showed good electrochemical performance for Li-ion batteries, and their performance is strongly affected by the gas composition used in the CO_x-OP process. For example, the first lithiation and delithiation capacities of Si-C prepared with CO_2-rich gas stream were 2481 mAh/g and 2061 mAh/g, respectively, resulting in high coulombic efficiency of 83.1% . In contrast, the first lithiation and delithiation capacities were 2148 mAh/g and 1356 mAh/g, respectively, resulting in low coulombic efficiency of 63.2.% for Si-C prepared with CO-rich gas stream.

Keywords:

Porous silicon-carbon (Si-C) composites, carbon oxide thermic oxidation (CO_x-OP) process, lithium-ion battery.

622570 Optimal Design of Controlled Environment Agricultural Systems Under Market Uncertainty

Abstract:

Global population growth and increased climate variability have posed major challenges to traditional agricultural production [2,6]. As these trends persist, meeting increased consumer demand under heightened uncertainty will further stress the capabilities of our existing food production infrastructure [4,5]. For this reason, alternative methods are being sought to augment conventional food production. Controlled environment agriculture (CEA) is an alternative strategy that decouples crop production from environmental conditions through indoor cultivation. CEA offers a more sustainable alternative to conventional agriculture and presents opportunities for improved food security by increasing public access to nutritious fresh produce. Startup costs for CEA systems, however, are significantly higher than those of traditional growing operations, and therefore present a larger barrier to adoption. To accelerate the adoption of this potentially disruptive technology for next generation food production, a better understanding of the long-term economic viability of CEA technology is desired. To provide this valuable insight, we have developed the first economics-based optimal design and planning model for CEA systems.

To date, the majority of agricultural planning models have focused on traditional outdoor food production. Many of the existing models have been tailored to highly specific operations, few have employed mathematically rigorous optimization techniques, and only a handful have accounted for uncertainty, which is desired to realistically assess the long-term economic viability of the system [1,3,7]. Since existing models are not applicable to CEA systems, we have developed an optimal design and planning model that can be used to assess the long-term economic viability of a broad range of CEA systems. Our proposed model employs robust optimization to account for uncertainty in the design and planning stages of CEA production, which enables risk-averse investment decision making for long-term economic viability.

In this work, we present a novel methodology for the simultaneous robust design and planning of CEA systems under multi-period market price uncertainty/risk. This problem is cast as a bilevel program with a nonconvex outer program pertaining to the design objective and multiple semidefinite inner programs pertaining to mean-variance risk exposure over each period considered in the planning horizon. The bilevel program is reformulated as a nonconvex semi-infinite program with several semi-infinite constraints pertaining to the multi-period risk exposure over the horizon. The problem is solved to global optimality using a custom branch-and-bound strategy constructed with EAGO.jl [10] that exploits a combination of McCormick-based envelopes and Gurobi's nonconvex solver [8], to solve the relaxed (upper-bounding in the maximization sense) subproblems, and IPOPT [9] for local solution of the lower-bounding (in the maximization sense) subproblems. A solution to the model represents a design and operating schedule that is robust to worst-case market uncertainty and therefore provides a conservative basis for CEA engineering and investment decision-making.

Our proposed model has been formulated generally and can be used to optimize large-scale CEA systems for the cultivation of any crop portfolio via single or multi-mode cultivation. The optimization objective employed is to maximize the net present value of the CEA system over its expected lifespan. Constraints limiting risk exposure and production capacity are also included to ensure that the optimal design is both robust and practical. Capital and operating expenses, including those associated with system construction, cultivation equipment, resources, labor, and distribution, are accounted for in the economic objective through sub-models that capture the location, capacity, and crop allocation cost dependencies of the specific CEA system being modeled. These sub-models can be easily adjusted for different CEA configurations without structural changes to the main optimization algorithm. This not only enables efficient determination of the optimal design for a given CEA configuration, but also enables efficient comparison between multiple candidate systems for strategic adoption of CEA technology.

To evaluate the efficacy of our approach, the economic performance of the robust design was compared against that of naïve design and operating strategies through two case studies. In each of the presented case studies, an economically feasible design was identified, even when the naïve design was economically infeasible, and the improvement in system performance attributed to the implementation of a robust design was quantified. This approach represents, to our knowledge, the first robust optimization approach to CEA systems and is demonstrated to effectively increase the robustness of CEA systems to market uncertainty, improve the long-term economics of CEA systems over naïve operating strategies, and validate the economic viability of single and multi-mode CEA production of distinct crop portfolios. This work establishes a means of evaluating the economic viability of a wide array of CEA systems and serves as a baseline for future development of CEA design and planning models to accelerate the adoption of CEA technology.

References:

[1] Ahumada, O., Villalobos, J.R., 2009. Application of planning models in the agri-food supply chain: a review. Eur. J. Oper. Res. 196 (1), 1–20. doi: 10.1016/j.ejor.2008.02.014 .

[2] Deschênes, O., Greenstone, M., 2007. The economic impacts of climate change: evidence from agricultural output and random fluctuations in weather. Am. Econ. Rev. 97 (1), 354–385. doi: 10.1257/aer.97.1.354 .

[3] Dury, J., Schaller, N., Garcia, F., Reynaud, A., Bergez, J.E., 2012. Models to support cropping plan and crop rotation decisions. A review. Agron. Sustain. Dev. 32 (2), 567–580. doi: 10.1007/s13593- 011- 0037- x .

[4] Fedoroff, N.V., 2015. Food in a future of 10 billion. Agric. Food Secur. 4 (1), 11. doi: 10.1186/s40066- 015- 0031- 7 .

[5] Field, C.B., Barros, V., Stocker, T.F., Dahe, Q., 2012. Managing the Risks of Extreme 1083 Events and Disasters to Advance Climate Change adaptation: Special Report of the Intergovernmental Panel on Climate Change (IPCC). Cambridge University Press.

[6] Food and Agriculture Organization of the United Nations (FAO), 2018. The state of agricultural commodity markets 2018: Agricultural Trade, Climate Change and Food Security. Technical Report. FAO. URL http://www.fao.org/3/I9542EN/i9542en.pdf .

[7] Glen, J.J., 1987. Mathematical models in farm planning: a survey. Oper. Res. 35 (5), 641–666. doi: 10.1287/opre.35.5.641 .

[8] Gurobi Optimization, LLC, 2020. Gurobi optimizer reference manual. URL http://www.gurobi.com .

[9] Wächter, A., Biegler, L.T., 2006. On the implementation of an interior-point filter line-search algorithm for large-scale nonlinear programming. Math. Program. 106 (1), 25–57. doi:

10.1007/s10107- 004- 0559- y.

[10] Wilhelm, M.E., Stuber, M.D., 2020. EAGO.jl: Easy advanced global optimization in Julia. Optim. Methods Softw. doi: 10.1080/10556788.2020.1786566 .

632232 Reproducible and Fast Mapping of Material Flows in Industrial Networks Using Piot Hub: A Novel Cloud Based Computational Tool

Abstract:

An important goal of industrial ecology is to ensure a balance is maintained between economic and environmental sustainability. Facilitation of this balance requires an understanding of the structure of the material economy and the material flows/waste generation in industrial networks in order to minimize the environmental consequences of these industrial systems. Physical Input-Output Tables (PIOTs) are uniquely suited for this purpose by mapping physical flows between industrial sectors. In previous work, PIOT-Hub was developed to automate the generation of both Physical Supply Tables (PSTs) and Physical Use Tables (PUTs) as well as their conversion into PIOTs by integrating the strength of mechanistic models and macroeconomic Input-Output framework [1]. Hence, PIOT-Hub, a cloud-based tool can be used to generate the PST and PUT for a region using a bottom-up approach where relevant industrial processes are simulated via mechanistic models to determine steady-state material input and output requirements. Thus, the tool establishes a connection between engineering and economic model in such a way that it is insightful for transitioning to a low carbon and zero waste circular economy. In past work, PIOT-Hub was applied to 11 agro-based economic sectors in Illinois, thus capturing elemental flows across the economy, demonstrating the automation aspect. Here we evaluate the reproducibility and time aspect of PIOT-Hub to map the material flows of equivalent economic sectors in Indiana. We develop simulation models for industries in Indiana and utilize PIOT-Hub for mapping the industrial network, thus evaluating the functionality of this novel computational platform. This work not only demonstrates the reproducibility of the tool's utilization to other geographic regions, but also serves as a case study of how the fast generation of PSTs, PUTs, and PIOTs by PIOT-Hub can be used to compare the material economies of multiple regions. We perform a comparative analysis of material flows in Illinois and Indiana using our work and the PIOT generated for Illinois in previous work. We will also demonstrate the collaborative aspect of this computational tool by live demo of the tool during this talk, thus highlighting the scaling impact of collaboration through cloud-based platform in making sustainable manufacturing decisions.

References

[1] Shweta Singh; Lan Zhao; Jaewoo Shin; Venkata Sai Gargeya Vunnava (2021), "PIOT hub for Automated Generation of Regional PIOTs," *https://mygeohub.org/resources/piotpy*.

625295 Solid Oxide Fuel Cell – Thermal Energy Storage (SOFC-TES) Integrated System Modeling and Energy Efficiency Optimization.

Abstract:

Concerns over greenhouse gas emissions and their associated environmental challenges continue to increase. As a result, cleaner energy generation technologies, such as fuel cells, are receiving attention as the next generation's energy sources. One of these attractive energy technologies is the solid oxide fuel cells (SOFCs) due to its several advantages such as 1) reduced greenhouse gas emissions, 2) higher efficiency, and 3) flexible fuel selection compared to other types of fuel cells. However, the temperature of flue gas from SOFCs is significantly high due to its operating temperature between 600°C and 1000°C. Therefore, it is necessary to recover this waste heat so that we can maximize the process energy efficiency.

Many studies have been conducted to utilize the high-temperature exhaust gases from SOFCs so far. However, they only focused on the direct utilization of the heat through the electricity generation, using processes such as gas turbine[1], Rankine cycle[2], and combined heat and power (CHP)[3] in order to maximize the process energy efficiency. However, the aforementioned utilization methods result in losses where the generated electricity is further converted to other forms of energy such as thermal energy for space heating. The energy losses during this process are often significant.

Motivated by this, we propose the SOFC-integrated thermal energy storage (TES) method as a more efficient utilization technique for recovering waste heat from SOFCs. We employed a 4 kW-SOFC and combined it with the thermal energy storage system for storing the excess energy from the high-temperature exhaust gas. In order to integrate both processes, SOFC and TES mathematical models were developed using MATLAB. Next, the developed models are integrated, simulated, and the overall efficiency of the SOFC-TES model is compared to the general utilization methods. To further improve the proposed model's thermal efficiency, we developed a heat exchanger network using Aspen Energy Analyzer. Through the developed heat exchanger network, the integrated model was optimized by pinch analysis.

The optimization results revealed that the process efficiency of the integrated system, that is, SOFC-TES, showed superiority compared with the traditional methods of utilizing the waste heat from SOFCs. By this, excess heat from the exhaust gases can be reliably stored during low-demand seasons and reutilized during high-demand seasons to supplement residential and industrial energy demand or deficits. We concluded that the SOFC-TES could be widely used for various purposes due to its inherent characteristics, i.e., storing heat, such as a greenhouse in the smart-farm and human life, compared with the traditional method of waste heat utilization from the SOFCs.

References

[1] S.H. Chan, H.K. Ho, Y. Tian, **Modeling of simple hybrid solid oxide fuel cell and gas turbine power plant**, Journal of Power Sources, 109 (1) (2002), pp. 111-120

[2] Masoud Rokni, **Thermodynamic analysis of an integrated solid oxide fuel cell cycle with a Rankine cycle**, Energy Conversion and Management, 51 (2010), pp. 2724-2732

[3] S.H. Chan, C.F. Low, O.L. Ding, **Energy and exergy analysis of simple solid-oxide fuel-cell power systems**, Journal of Power Sources, 103 (2002), pp. 188-200

630767 Upcycling Food Waste for Hydrogen Storage: Technoeconomic Assessment

Abstract:

While annually 1.6 billion tons of food waste is estimated to be discarded globally by 2030, according to U.S. Environmental Protection Agency (EPA) the corresponding figure is projected forecast to reach approximately 66 million tons for U.S. With the monumental challenge of food waste management, researchers are seeking avenues of upcycling FW into higher value materials such as hydrogen storage medium. Therefore, the objective of this research was to evaluate the technoeconomic prospect of activated carbon production from food waste for the ultimate purpose of hydrogen storage through the application of hydrothermal carbonization of food waste followed by chemical activation of the hydrochar using KOH. The process was simulated using Aspen Plus and the process flow diagram (PFD) was constructed such that it mainly constituted of 2 rotary kilns and filter press, 1 rotary cooler, dryer, homogenizer, as well as other necessary equipment - pumps, heat exchangers, storage tanks. The facility processes 96 ton of food waste per day to be converted to activated carbon powder with a carbon yield of 20.4%. Also, it was assumed that the activated carbon manufacturing facility obtains food waste of Brevard County, Florida and the facility operates continuously 330 days a year. Several parameters were incorporated in the study including equipment sizing, capital costs and operating costs, such as labor, utilities, maintenance and equipment depreciation. Break-even cost from the economic analysis was evaluated and sensitivity analysis was carried out to reveal the effect of 7 different process variables on the break-even cost where raw materials' (KOH and food waste) and wastewater treatment cost significantly affected it. The unprecedented analysis shed light on the economic feasibility of producing activated carbon from food waste as a cost-effective route of upcycling it.

625694 Economic and Environmental Impact of Biochar Addition on Anaerobic Digestion Process

Abstract:

Anaerobic digestion (AD) technology is becoming a promising pathway for producing clean and renewable natural gas from waste resources. Despite growing industrial adoption, AD technologies still suffer from inefficiency due to process instability and microbial inhibitions. Scientists have demonstrated that biochar addition has positive impacts on AD. Biochar addition increases biogas yield and methane concentration and reduces gas contaminants. This influence is attributed to several mechanisms: providing support for microbial colonization, adsorbing excess inhibitory products, and improving electron transfer between syntrophic microorganisms. However, due to the cost and energy-intensive conversion process, the additional costs for biochar addition could offset the resulting energy benefits. This study investigates an integrated biorefinery where corn stover biochar is added to the AD of animal manure to produce biogas-derived Renewable Natural Gas (RNG) for pipeline injection.

A kinetic model of AD, biogas upgrading, and pipeline injection is simulated in Aspen Plus ™ for a mid-size farm with 5000 cattle heads. The process simulation model estimates a 28% increase in biogas yield for 3% (wt. % of wet mass) of biochar addition. High pressurized water scrubbing process is used for biogas upgrading into RNG. Techno-economic analysis (TEA) determines the minimum RNG selling price for a 30-year plant lifetime with a 10% Internal Rate of Return (IRR). TEA estimates the minimum RNG selling price as $42/MMBTU and $35/MMBTU for the without biochar and with biochar scenario, respectively. Biochar addition lowers the RNG selling price by 16% whereas adding revenue from digestate selling and renewable identification number (RIN) credits result in up to 32% reduction. Biochar addition increases the capital and operating costs by 8% and 13%, respectively. Cradle-to-gate life cycle analysis (LCA) is performed in SimaPro ™ to evaluate the GHG emissions of biochar amended AD for RNG production. GHG abatement for the without biochar and with biochar scenario is 0.038 kg $CO_{2\ eq}$/MJ of RNG and 0.123 $CO_{2\ eq}$ /MJ of RNG, respectively. LCA results show biochar addition leads to 2.18 times GHGs abatement compared to the without biochar scenario. Environmental benefits mostly come from the additional carbon sequestration with biochar land use and fossil fuel displacement by syngas from pyrolysis. Sensitivity analysis shows the influence of different major processes and economic parameters on the RNG selling price. The process is cost-prohibitive for small-scale farms due to the significant economy of scale penalty for the biogas upgrading and RNG injection facility. Increased biogas yield reduces the manure input leading to less digestate production. This would decrease the environmental benefits of emissions avoidance from conventional manure management and carbon sequestration per MJ of RNG while reducing the emissions from digestate transportation and applications. Uncertainty analysis considers the variations of key technical and economic parameters within a reasonable range and gives an estimated range of fuel selling prices and GHG abatement. Minimum RNG selling prices vary between $8.82/MMBTU and $16.28/MMBTU and GHG abatement vary between 0.06 kg $CO_{2\ eq}$/MJ and 0.16 kg $CO_{2\ eq}$/MJ within 10-90% confidence intervals.

627985 Optimizing Syngas Production Process By Downdraft Gasifier Using Response Surface Methodology

Abstract:

The gasification process of biomass to produce combustible gases has an attracted interest recently. The downdraft gasifier has an advantage over other types of gasifiers because of its small size, high quality gas production, and low tar yield. One of the challenges in the gasification process is the production of low-quality syngas (low concentration of H2 and CO) and large amount of tar. To overcome this problem, optimization of the gasifier operating conditions was employed to achieve high-quality product gases and hence high gasification process efficiency. In this work, the influence of initial gas temperature, airflow rate, and mole fraction of oxygen in primary air, on the gasification efficiency and product quality was optimized by response surface methodology. The obtained results showed that the reduction in the temperature caused a low concentration of carbon monoxide and a high amount of tar in the product, thereby there was a decline in the efficiency of the gasification process.

623925 Techno-Economic Evaluation of Nano-Porous Silica Production from Rice Husk and Sand

Abstract:

Global rice production in 2019 was estimated at about 500 million tons. Rice husk is a by-product from rice, including about 20 wt% of rice and is needed to be disposed as waste. Rice husk contains approximately 14-25% of ash. The silica content in rice husk ash (RHA) is above 90%. Nanoporous silica (NPS) used in industrial fields such as reinforced rubber additives (tires), concrete mixture (cement binder), nano-fibers, nano-filters, and membranescan be produced from RHA. Rice hisk is combusted at a relatively low temperature (650-800°C) and RHA is produced, whereas the silica sand is operated at 1400-1500°C to produce silicate. In addition, compared to the conventional NPS production from sand, The NPS production from RHA has an advantage in lowering the production cost due to low material cost and energy consumption.

In this study, the techno-economic feasibility for the sodium silicate ($Na_2O{\cdot}xSiO_2$)-based nano-porous silica (SS-NPS) production from sand and rice husk was performed. The return on investment (ROI) and payback period (PBP) were evaluated according to the plant size for the two cases. The effects of raw material and product prices on ROI and PBP were analyzed and economic conditions to be economically viable were suggested.

AIChE

The Global Home of Chemical Engineers

A CAPITAL AND ENERGY EFFICIENT ALTERNATIVE EMISSIONS CONTROL TECHNOLOGY FOR CONTROLLING EMISSIONS USING BEAD ACTIVATED CARBON

Kim K. Tutin, Captis Aire LLC, East Point, GA United States
Harold R. Cowles, Environmental C&C Inc., Scottsdale, AZ United States
John M. Berger P.E., Environmental C&C Inc., Fishers, IN United States

TABLE OF CONTENTS:

I. INTRODUCTION:

In manufacturing, process exhaust gas (air) streams released may contain volatile organic compounds (VOCs), Hazardous Air Pollutants (HAPs), particulate, etc. These are considered pollutants/emissions and must be removed from the air prior to release to avoid air pollution emissions into our communities. Add-on controls either collect the VOCs for recovery and reuse or destroy the VOCs. A number of control options are available including thermal oxidizers, catalytic oxidizers, and biofiltration which destroy pollutants. Other control systems including condensation, scrubbing, and adsorption systems may recover pollutants.

This paper provides an introduction to an adsorption technology, the Fluidized Bed Concentrator (FBC) with Bead Activated Carbon (BAC) media. It describes 1) the FBC technology, 2) the BAC media used in these systems, 3) the performance characteristics, 3) typical applications of this technology, and 4) a case study review of the FBC technology demonstration for controlling emissions and collecting terpenes from wood drying.

II. DESCRIPTION OF FBC TECHNOLOGY: FOUR PRIMARY PARTS

1. Adsorber:

The adsorber section has trays filled with very adsorbent BAC media. Contaminated process exhaust enters from the bottom, passes upward through the adsorption trays, and fluidizes the Bead Activated Carbon (BAC) adsorbent media. When the process air interacts with the BAC, it adsorbs (captures) the terpenes Volatile Organic Compounds (VOCs) in its pores. VOC saturated BAC flows to the bottom of the adsorber vessel and is transferred using fans to pneumatically transfer it to the desorber. Cleaned (reusable) BAC from the desorber feeds continuously into the top of the adsorber, providing counter-current VOC removal. Cleaned air exits out the top of the adsorber.

2. Desorber:

In the desorber, the BAC media is heated (regenerated), causing it to release the VOC vapors into a low volume, gas stream. This gas stream is then circulated continuously to the next stage in the process. The regenerated (cleaned) BAC is then returned continuously to top of the adsorber for reuse. Returning this regenerated BAC to the top of the adsorber provides a final "polishing" of emissions to remove residual emissions prior to release of the process exhaust air.

"All Electric" Option

Desorber Recovery System: In the recovery system, a low volume inert carrier (nitrogen) gas is introduced into the bottom of the desorber. Gas flows up through the desorber as the BAC flows downward. The system is heated via either hot oil or electric heating. The heat causes the VOCs to volatilize off the BAC. The VOC vapors are liberated resulting in a concentrated waste flue gas stream that moves from the desorber to the condenser. This system may eliminate the on-site use of natural gas entirely. Thus, it offers an "All Electric" control technology.

Desorber Oxidation System: In the oxidation system, clean hot stripping air at 100-700°F is introduced to the bottom of the desorber. It flows upward across the BAC media to both volatilize the VOCs and fluidize the BAC. The VOC vapors are liberated resulting in a concentrated waste flue gas stream that moves from the desorber to the oxidizer.

3. Condenser or Oxidizer:

The gas stream from the desorber is circulated continuously to this system for further treatment to either capture and recover VOCs or oxidize them.

"All Electric" Option

Condenser Recovery System: In the recovery system, nitrogen gas circulates from the desorber to the condenser continuously. The VOC vapors are sufficiently concentrated such that the vapors can be cooled via a tube and shell condenser converting the vapor to liquid. The liquid drains continuously into a storage system. The liquid VOCs can then be reused or sold. The nitrogen recirculates back to the desorber where it entrains the VOCs vapors again and the cycle continues.

Oxidizer Oxidation System: In the oxidation system, the vapors are instead sent to a small thermal oxidizer to treat the VOCs. While this system does not provide recovered (collected) VOCs, it nevertheless retains FBC technology benefits including significantly reducing natural gas usage. The burner component of the thermal oxidizer is selected to provide enough energy to increase the thermal oxidizer to the 1400°F operating temperature. The burner is designed to minimize the "fire rate and thus consume negligible fuel. Optionally, a "low NO_x" design can be selected to minimize production of NO and NO_2 (oxides of nitrogen). Thus, the system minimizes energy usage and GHG emissions. [1]

4. Side Stream Reactivator (SSR) [2]:

A small portion of the BAC media exiting the desorber may be transferred to an optional SSR. The SSR reactivates the BAC by heating it to 1450°F with the introduction of a small amount of water which reactivates the BAC continuously to regenerate it and thus maintain its ability to adsorb VOCs. Reactivated BAC from the SSR is transferred back to the top of the adsorber for reuse.

III. ADSORBENT MEDIA:

The BAC adsorbent media utilized in this system is an essential element of the FBC technology. Specialized Bead Activated Carbon (BAC) efficiently adsorbs VOCs (i.e. terpenes) from the process air. The BAC has high-fill (adsorption) capacity, is a small sphere-shaped particle that is less than 1 mm in diameter with a narrow particle size distribution, has low dust, has high purity, has high flowability, and has high strength and wear resistance as measured by ASTM D 3802 – 79 (2005) Standard Test Method for Ball-Pan Hardness of Activated Carbon.[3] The spherical shape enables the BAC to fluidize and flow.

The BAC media can be reactivated and reused. This is accomplished either by either continuous or batch processes.

---

[1] "VOC Removal and Destruction System and Method". U.S. Patent and Trademark Office. (USPTO) https://appft.uspto.gov/netacgi/nph-Parser?Sect1=PTO2&Sect2=HITOFF&p=1&u=%2Fnetahtml%2FPTO%2Fsearch-bool.html&r=1&f=G&l=50&co1=AND&d=PG01&s1=20190022579.PGNR.&OS=DN/20190022579&RS=DN/20190022579 (Accessed October 4, 2021).

[2] "GASTAK® Solvent Recovery and Odor Removal by Fluidized Bed Process". Kureha Ecology Management Co. , LTD website. https://www.kurekan.co.jp/en/download/gastak.pdf (Accessed September 18, 2021).

[3] "ASTM D3802-79 (2005) Standard Test Method For Ball-Pan Hardness Of Activated Carbon. https://webstore.ansi.org/standards/astm/astmd3802792005 (Accessed September 18, 2021).

Continuous Reactivation: In the continuous process, a small amount of BAC media is diverted from the BAC exiting the desorber and sent to an in-line side stream reactivator (SSR).[4] It is then returned to the top of adsorber. Returning this reactivated BAC to the top of the adsorber provides a final "polishing" of emissions to remove residual emissions prior to release of the process air.

Batch Reactivation: In the batch process, for operators that do not have an SSR, the BAC media is unloaded periodically, shipped offsite for reactivation, and then reloaded once returned. In the interim, a second load of BAC is utilized. Reactivation is typically performed by California Carbon who has been providing reactivation services since 1975.[5] This service includes a reactivation warranty *"Carbon will be reactivated to 95% of Virgin Activity per ASTM D 3467 and Carbon will lose no more than 10% of incoming volume of the specified mesh size in the reactivation process."*[6] BAC lifetime is site and application specific. Some operators have been reusing their BAC for years.

Bead Activated Carbon (BAC) Adsorbent Media

Bead Activated Carbon (BAC), the Adsorbent Media, Fluidized in Tray

IV. FBC OPTIONS: RECOVERY AND OXIDATION

The Fluidized Bed Concentrator can be utilized to either recover the VOCs or oxidize them as shown in the figures. Operators typically select the desired option based on a number of factors especially including the anticipated value and reusability of the recovered VOCs. The value of the VOCs depends on a number of factors including the anticipated quantity recovered, purity, salability, and/or reusability.

Recovery: When the value of the recovered VOCS is high and if the cost of recovery is less than the cost of purchasing new material, then recovery makes sense.

---

[4] "GASTAK® Solvent Recovery and Odor Removal by Fluidized Bed Process". Kureha Ecology Management Co. , LTD website. https://www.kurekan.co.jp/en/download/gastak.pdf (Accessed September 18, 2021).

[5] "California Carbon Co. Inc. Welcome". California Carbon website. http://www.californiacarbon.com/index.html (Accessed September 18, 2021).

[6] "Custom Spent Carbon Reactivation". California Carbon website. http://www.californiacarbon.com/pdfs/Custom%20Reactivation.pdf (Accessed September 18, 2021).

Oxidation: If the VOC has low or no recovery value, as, for example, when it is a mixture, or if it poses disposal concerns, such as for toxic compounds, then destruction probably makes the most sense. Excluding value derived from VOC recovery, the oxidation option retains FBC system performance characteristics including reduction in energy usage.

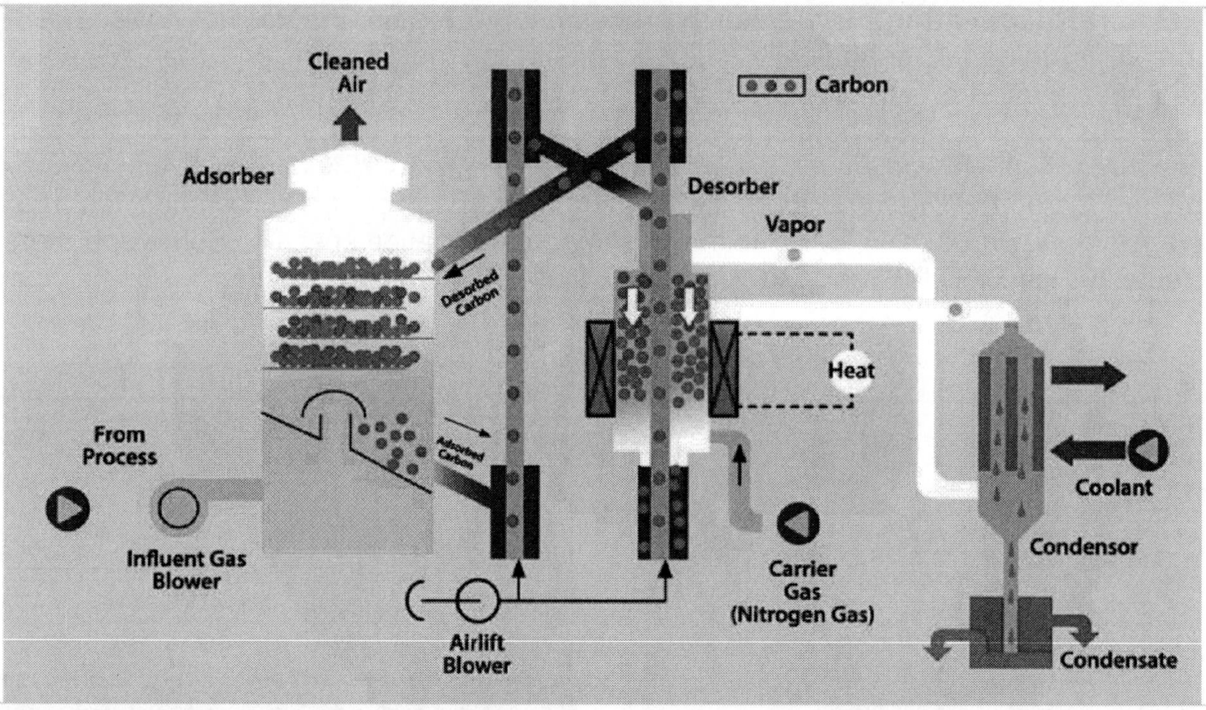

RECOVERY SYSTEM: Adsorbs, Desorbs, and Condenses VOCs

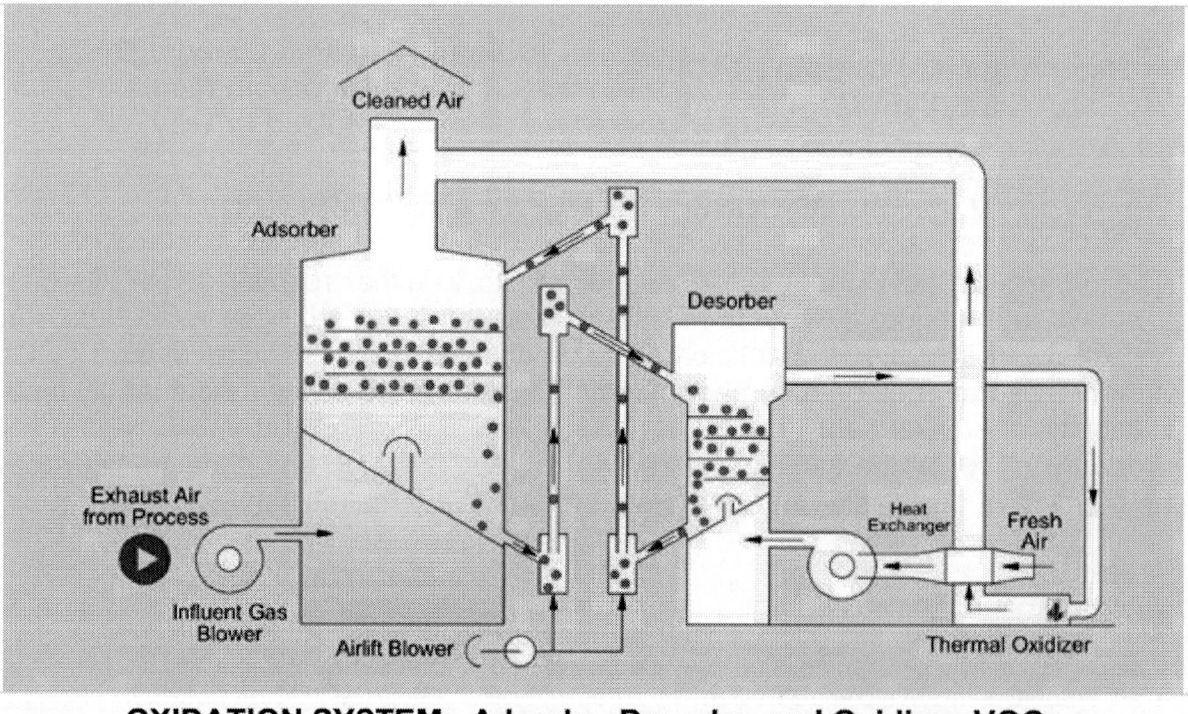

OXIDATION SYSTEM: Adsorbs, Desorbs, and Oxidizes VOCs

V. PROCESS PERFORMANCE CHARACTERISTICS:

1. The FBC reduces energy usage (i.e. natural gas) versus oxidation systems due to reduction or elimination of energy used to heat the air going through thermal oxidizers:

Baseline Oxidation Systems:

Thermal oxidizers burn VOCs by heating them in enclosed combustion chambers in the presence of excess oxygen. The precise temperature and residence time needed for effective reduction of VOC emissions varies with the nature of the VOC. However, temperatures of 1400-2000°F and residence times of 0.5-2 seconds are typical. Since auxiliary fuel is necessary to maintain combustion temperatures, most thermal oxidizers incorporate some form of heat recovery to lower auxiliary fuel costs.

FBC Technology:

The air going through the adsorber is not heated. Thus, the natural gas usage is reduced or eliminated.

Recovery – "All Electric" Option: For a recovery system, the control device can be "All Electric". It could eliminate the use of natural gas entirely.

Oxidation: The FBC oxidation system significantly reduces natural gas usage but does not eliminate it.

Energy Reduction Calculation: The amount of energy that is eliminated, that would have been used by the thermal oxidizer, can be calculated utilizing one of many credible sources. One of the best sources is the Institute of Clean Air Companies' Guidance Methodology[7]. Calculators are also available on-line including those provided by the EPA[8], Cycle Therm[9], and Captis Aire[10]. With these calculators, users can input their specific data and the calculators will calculate the estimated amount of energy required to run the thermal oxidizer. All these sources rely on a single primary factor to calculate energy required to heat the air. This factor is the specific heat of air which is the number of British Thermal Units (BTUs) required to heat one pound of air by one °F. This factor along with the system specific factors including the standard cubic feet per minute (sCFM) of air, the temperature increase, the density of air, and conversion factors provide the number of BTUs required to heat the air going through the device. These factors are used together with the energy efficiency and the heat generated by the oxidation of the emissions to calculate the BTUs of energy required to heat the air going through the RTO. Users can then multiply BTUs times the forecasted long-term cost of natural gas to estimate the long-term cost of natural gas required over the 20+ year life of the thermal oxidizer. This calculation provides the value of the natural gas elimination.

---

[7] "Institute of Clean Air Companies (ICAC) Guidance Method for Estimation of Gas Consumption in a RTO". Institute of Clean Air Companies website. https://cdn.ymaws.com/www.icac.com/resource/resmgr/RTO-F1.pdf (Accessed September 18, 2021)

[8] "Air Pollution Control Cost Estimation Spreadsheet". Environmental Protection Agency (EPA) website. https://www.epa.gov/sites/production/files/2018-01/oxidizers_calc_sheet_finalversion_1-16-2018.xlsm (Accessed Feb 3, 2021).

[9] "Operating Cost Calculator". Cycle Therm website. http://www.cycletherm.com/resources/operating-cost-calculator.aspx (Accessed Feb 3, 2021).

[10] "Calculate Engineering Estimate of Natural Gas Usage and Greenhouse Gas Generation". Captis Aire website. https://www.captisaire.com/Value-Calculator/ (Accessed Feb 3, 2021).

2. <u>The FBC generates revenue from sales of recovered organics:</u>

When using a recovery system, including the FBC, operators have the option to collect and sell or reuse the volatile organic compounds recovered. The liquid VOCs are collected from the FBC condenser for reuse or sale. The value of recovered VOCs will depend on the type, amount, and purity of VOCs collected.

3. <u>The FBC reduces Greenhouse Gases (GHG) emissions versus oxidation systems due to the reduction or elimination of burning natural gas:</u>

Due to the reduction or elimination of natural gas usage and reduction in electrical usage by the FBC, greenhouse gas emissions are also reduced. The greenhouse gas emissions reduction can be calculated based on the amount of carbon in natural gas eliminated and its stoichiometric conversion to carbon dioxide (CO_2). The Environmental Protection Agency (EPA) has an on-line calculator[11] that enables the user to calculate greenhouse gas emissions that are eliminated by eliminating natural gas used to heat the air in the thermal oxidizer.

4. <u>The FBC enables creation of carbon offsets versus oxidation systems:</u>

In conversations with American Carbon Registry[12] and ClimeCo[13], it may be possible to measure, register, and verify carbon offsets for a site converting to the FBC technology from an oxidation system due to the reduction in energy usage.[14]

The quantification methodology and value of carbon offsets is evolving quickly. But these carbon offsets can be sold or traded externally. Or they may be utilized for internal credit accounting purposes. Potential values for these carbon offsets vary widely.[15] However, as of early 2021, American Carbon Registry and ClimeCo estimated the value of carbon offsets for an FBC project at $5/ton.[16]

5. <u>The FBC reduces electricity usage versus oxidation systems:</u>

The horsepower required to run the fans to move the process exhaust air through the BAC media is reduced for the FBC due to the fact that it's BAC is fluidized (moving) rather than packed (non-moving). The pressure drop across the system is reduced from a typical 10-20" to 3.5". Thus, the horsepower required to run the fans for the FBC may be ~ ¼ of that required for the RTO.

To estimate the electrical savings, the sum of electrical demands from the FBC is subtracted from the sum of electrical demands of the thermal oxidizer. Net kW·h along with site-specific electricity costs and operating hours per year are used in combination to estimate electrical savings.

---

[11] "Greenhouse Gases Equivalencies Calculator - Calculations and References". EPA website. https://www.epa.gov/energy/greenhouse-gases-equivalencies-calculator-calculations-and-references (Accessed December 28, 2020).

[12] "Harnessing the Power of the Markets to Improve the Environment". American Carbon Registry website. https://americancarbonregistry.org/ (Accessed February 3, 2021).

[13] "Scalable greenhouse gas projects to earn real financial returns for your business." ClimeCo website. https://climeco.com/ (Accessed February 3, 2021).

[14] "AM0044: Energy efficiency improvement projects - boiler rehabilitation or replacement in industrial and district heating sectors --- Version 2.0.0". Clean Development Mechanism (CDM) website. https://cdm.unfccc.int/methodologies/DB/3HZ4USHZ2W449HMAXZN420E5PJB1QF (Accessed Sept 18, 2021)

[15] "Allowance Price Explorer". International Carbon Action Partnership website. https://icapcarbonaction.com/en/ets-prices (Accessed Jan 20, 2021).

[16] Eric Ripley (Director Industrial Programs, American Carbon Registry), in a phone conversation with Kim Tutin November 17, 2020.

6. **The FBC enables superior mass transfer of VOCs through counter current flow fluidized BAC media that is continuously regenerated:**

Baseline Adsorption Systems:

Adsorption: In sorbent treatment systems, process exhaust air moves across the sorbent. VOCs are removed via attraction (adsorption) of the VOCs on carbon or zeolite sorbent media in packed (non-moving) beds. This may be accomplished by concentrators that have wheels with sectors containing activated carbon or zeolite media. In this arrangement, most of the space on the wheel functions in adsorption mode, with one small sector being regenerated at any given time. Or the process exhaust air may be forced through large vessels containing sorbent media. In this arrangement, the VOCs are treated via a once-through process.

Regeneration / Reactivation: Regeneration is accomplished by heating the sorbent media with heaters, heated gases, steam, etc. If using steam, water may be a byproduct. This can be accomplished either via off-site regeneration or by alternating the flow between multiple vessels where one vessel functions in adsorption mode while the other is being regenerated at any given time. In both these arrangements, the sorbent media is being regenerated in batch or semi-continuous fashion. Reactivation may be accomplished by sending the sorbent media off site for reactivation. Or the sorbent media may be discarded when it becomes saturated.

Effect of water vapor: Especially for process exhaust air heavily laden with water vapor, i.e. 10% B_{ws} moisture by volume[17] or more, the sorbent media may become partially or fully saturated with water. At high moisture content, water molecules begin to compete with the hydrocarbon molecules for active adsorption sites. This reduces the capacity and the efficiency of the adsorption system.

FBC Technology:

Adsorption: The BAC media in the fluidized bed concentrator adsorber is fluidized (moving) by the process air moving opposite (counter current) to the BAC as it moves downward through the trays in the adsorber. This provides excellent mixing of the BAC and contact with the air. As this occurs, VOCs or odors are transferred from the process air to the BAC. The cleaned air exits through the top of the adsorber.

Regeneration / Reactivation: The BAC media is regenerated continuously via continuous recirculation to and from the desorber. It is reactivated continuously if system includes the Side Stream Reactivator (SSR). This ensures the BAC retains the capacity to maintain adsorption efficiency. This is done at temperatures well above the boiling point of water, often around 400°F or more. This minimizes or eliminates accumulation of water and/or organics on the BAC.

Effect of water vapor: Continuous regeneration and reactivation at temperatures above the boiling point of water, i.e. 400°F, volatilizes the water off the BAC media and prevents the BAC from becoming saturated with water. Thus, it maintains adsorption efficiency even when treating process exhaust streams with high moisture contents, as are typical for wood drying applications, i.e. ≥20% B_{ws} moisture by volume[18]. This system minimizes or eliminates need for dehumidification systems.

---

[17] "METHOD 4—DETERMINATION OF MOISTURE CONTENT IN STACK GASES". EPA website. https://www.epa.gov/sites/default/files/2016-06/documents/m-04.pdf. (Accessed October 17, 2021).
[18] "METHOD 4—DETERMINATION OF MOISTURE CONTENT IN STACK GASES". EPA website. https://www.epa.gov/sites/default/files/2016-06/documents/m-04.pdf. (Accessed October 17, 2021).

Effect of Fluidized (moving) vs Packed (non-moving) media: BAC media fluidization prevents pockets of high concentration VOCs, provides higher dry particulate tolerance (reduced plugging of interstitial spaces between media), and enables continuous on-site regeneration and reactivation.

TABLE I
PHYSICAL AND OPERATIONAL COMPARISON OF THE FBC VS THE RTO

Parameters:	RTO	FBC	Advantages of FBC
Treatment Temperature, °F	>1400°F	175°F	• Adsorption at process air temperature • No natural gas required for heating
Media Desorption Temperature, °F	>1400°F	400°F	• Much less energy required for desorption • Continuous media regeneration, above the boiling point of water, prevents the media from getting saturated with water. Thus, it maintains its ability to adsorb pollutants.
Media depth, feet (ft)	16 ft	1 ft	• Less Electricity (fan HP) required to move air through less media
Media Configuration	Packed (Not Moving)	Fluidized (Moving)	• Less Electricity (fan HP) required to move process air through fluidized media
Pressure Drop, Inches (")	10-20"	3.5"	• Less Electricity required to move process air against less pressure
Fan size, Horsepower, (HP)	2400 HP	600 HP	• Less HP due to media that is fluidized. • Less susceptible to plugging from particles
Emissions Treatment	Oxidation to CO^2	Adsorption (Capture)	• Emissions are NOT oxidized to CO_2, so reduces GHG emissions • Emissions collected can be sold or reused
Media Reuse	Discarded	Reused	• Media desorption with 'in unit' continuous regeneration. • Flowable bead shaped media can be removed and reused
Safety		Inherently Safer	• Lower operating temperatures • No large heavy moving parts • Easier cleaning and maintenance

VI. AWARDS:

The FBC has been recognized for excellence in environmental performance. Honda Motors of Alabama was recognized as *"Air Conservationist of the Year"* when they implemented the FBC to control emissions when they increased production.[19] Ford's "Fumes to Fuels" work was awarded the EPA's *"Clean Air Excellence"* award.[20] In this work, they utilized the FBC to capture solvent-laden air, strip the solvent from the air, concentrate the solvent up to 2,000 times, and send it to a fuel reformer. The fuel reformer used the concentrate to generate hydrogen-rich fuel, which was directed into a fuel cell to generate electricity.

VII. APPLICATIONS:

Concentration systems, including the FBC, are advantageously used in front of control devices when treating very high volumes of very dilute exhaust gas to reduce energy usage. Recovery systems are beneficial when the value of the recovered VOCs is high. The following are a few examples where the FBC technology has been utilized historically.

TABLE II
FBC APPLICATIONS

Application	Chemistry
Adhesive Tape Mfg	Toluene
Agricultural Products	Organic Acid & Ammonia Odor
Aluminum Casting	Phenol, Formaldehyde, Ammonia
Animal Lab	Odor Control
Atomic Power Plant	Styrene
Auto Parts Degreasing	Trichloroethylene
Bakery	Odor Control
Brewery	Odor Control
Car Wash	Kerosene
Cellophane Coating	Toluene, Ethyl Acetate, Butyl Acetate
Ceramic Condenser	Terpineol
Confectionery	Odor control
Chemical Production	Toluene, Higher Molecular Weight Alcohols, Carbon Tetrachloride

19 "HMA Honored as Air Conservationist of the Year". Honda website. https://www.hondaalabama.com/article/hma-honored-as-air-conservationist-of-the-year (Accessed September 21, 2021).

20 "Clean Air Excellence Award Recipients". Environmental Protection Agency (EPA) website. https://www.epa.gov/sites/default/files/2015-06/documents/clean_air_excellence_award_recipients_year_2003.pdf (Accessed September 21, 2021).

TABLE II (CONTINUED)
FBC APPLICATIONS

Application	Chemistry
Dye Production	Perchloroethylene
Electronics Mfg Plant	DMF, MEK, etc.
Electronics Parts Cleaning	Trichloroethylene, Toluene, 1,1,1-Trichloroethane
Fabric Washing	Perchloroethylene
Feed Processing	Odor Control
Film Coating	MEK, Methyl Cellosolve, Toluene, Tetrahydrofuran
Film Laminating	Toluene, Ethyl Acetate, n-Hexane
Gravure Printing	Toluene, Xylene, n-Propanol, n-Propyl Acetate, mixed
Iron Casting	Phenol, Formaldehyde, Ammonia
Lacquering of Film	Toluene, Ethyl Acetate, MEK
Landfill	Odor Control
LSI Mfg Plant	Phenol, Acetone, Methanol, Mixed Solvents
Lubrication Mfg	Mixed Solvent Odor Control Magnetic Tape, Chlorinated Hydrocarbon
Medicine Mfg Plant	Butanol, Methylene Chloride
Paint Booth	Thinner
Printing Ink Mfg	Mixed Solvents
PVC Resin Mfg Plant	Vinyl Chloride
Resin Mfg	N-Hexane, Methylene Chloride
Rubber Coating	Toluene
Rubber Vulcanizing	Odor Control
Sandpaper Mfg	Toluene, Xylene, Ethyl Acetate
Semiconductor Mfg	IPA, HMDS, NMP, Ethanol, Methanol, Ethyl lactate
Silk Screen Printing	Xylene, Mixed Solvents
Teflon Tubing Mfg	Perchloroethylene
Wastewater Aeration	Odor Control

TABLE III
EXAMPLES OF FBC SYSTEMS INSTALLED

FLUIDIZED BED CONCENTRATOR SYSTEMS INSTALLED: [21]	
Industry **Location** **Inlet Stream** **VOCs** **Performance Achieved** **Notes**	Semiconductor Chip Manufacturing Israel 20,000 CFM Semiconductor solvents, 50 ppmv 95% Reduction Efficiency System located in populated area of a major city has eliminated nuisance odor
Industry **Location** **Inlet Stream** **VOCs** **Performance Achieved** **Notes**	Semiconductor Chip Manufacturing Arizona 36,000 CFM Semiconductor solvents, 25-100 ppmv 90%+ Reduction Efficiency System met strict local compliance requirements while achieving Zero Unscheduled Downtime (ZUD)
Industry **Location** **Inlet Stream** **VOCs** **Performance Achieved** **Notes**	Teflon Tubing Manufacturing Massachusetts 1,800 CFM Perchloroethylene, 1500 ppmv 98% Reduction Efficiency Recovered solvent re-used in process. System payback in 1 year.
Industry **Location** **Inlet Stream** **VOCs** **Performance Achieved** **Notes**	Furniture Painting California 50,000 CFM Paint Solvents, 25-50 ppmv 90%+ Reduction Efficiency Recovered solvent and reformulated with paint
Industry **Location** **Inlet Stream** **VOCs** **Performance Achieved** **Notes**	Military Vehicles Michigan 150,000 CFM Paint Solvents 95% Reduction Efficiency Controlled multiple different interchangeable VOC sources.

[21] "Environmental C&C Inc". Environmental C&C website. https://www.environmentalcc.com/ (Accessed September 18, 2021).

Automotive Site, Alabama, 100,000 CFM
Paint Solvents, >90% Reduction Efficiency
Two adsorbers and one desorber

Image shows three adsorbers one desorber

Image shows ductwork that enabled control of any of the 3 sources to any of the 3 adsorbers.

Each adsorber has its own process fan.

System provided control of multiple different interchangeable VOC sources

Military Vehicles, Michigan, 150,000 CFM
Paint Solvents, 95% Reduction Efficiency

TABLE IV: VALUE CREATION CALCULATION EXAMPLE

Fluidized Bed Concentrator (FBC) vs Regenerative Thermal Oxidizer (RTO)	Value per Year
1) Reduce Energy Usage (Natural Gas) due to Elimination of Energy Used to Heat the Air in the RTO: [22] [23] a. $\frac{mmBTU}{Year}\ to\ heat\ air\ in\ RTO = (\Delta\, \mathrm{T})\ (sCFM)(1.08)(\%Heat\ Loss)(\frac{Operating\ Hours}{Year})/1000000$ b. $(1500°F - 175°F) * (300{,}000sCFM) * \left(\frac{1.08\ BTUs}{sCFM \cdot Hour \cdot Degree°F}\right) * (7.5\%) * (8000\frac{Hrs}{Yr})/1000000$ c. $= \left(257{,}580\frac{mmBTU}{Yr}\right) - [\left(250\frac{lbs\ VOC}{Hour}\right) * \left(15{,}000\frac{BTUs}{lb\ VOCs}\right) * \left(\frac{8000\ Hrs}{Year}\right)] = \mathbf{227,580}\frac{mmBTUs}{Year}$ d. $\left(\frac{227{,}580\ mmBTUs}{Year}\right)(\$3.5\frac{Long\ Term\ Cost\ of\ Natural\ Gas}{mmBTU}) = \mathbf{\$796,530\ Savings/Year}$	Save by Eliminating Natural Gas **$796,530**
2) Reduce Greenhouse Gases (GHG) Generated from Burning Natural Gas: [24] a. $Tons\ GHG = (mmBTU) * \left(\frac{117\ lbs\ CO2}{mmBTU}\right) * (\frac{1\ Metric\ Ton}{2205\ lbs})$ b. $\left(227{,}580\frac{mmBTUs}{Yr}\right) * \left(117\ lbs\frac{CO2}{mmBTU}\right) * \left(1\frac{Metric\ Ton}{2205\ lbs}\right) = \mathbf{12,076}\frac{Tons\ GHG\ Reduction}{Year}$	Sell Carbon Offsets **$60,378**
3) Sell Carbon Offsets (Carbon Credits) Utilizing Carbon Accounting Service Providers, i.e. ClimeCo: [25] a. $\left(\frac{12{,}076\ Tons\ GHG}{Year}\right) * \left(\frac{\$5}{Ton\ GHG}\right) = \mathbf{\$60,378/Year\ Selling\ Carbon\ Offsets}$	
4) Reduce Electricity Usage by Reducing Fan Power to Move Process Air through Fluidized Media: $(2400RTO - 600FBC\ HP\ fans - 800\ HP) * \left(\frac{0.75\ kW \cdot h}{HP}\right)\left(\frac{\$0.07}{kW \cdot h}\right)\left(\frac{8000\ Hrs}{Year}\right) = \mathbf{\$420,000}\frac{\mathbf{Savings}}{\mathbf{Year}}$	Save by Reducing Electrical Usage **$420,000**
5) Sell Terpenes [26] **Yielded from Green Pine Wood** [27]**:** a. $\left(\frac{1\ Gallon\ Terpenes}{Oven\ Dry\ Ton\ Wood}\right) * \left(\frac{300{,}0000\ Dry\ Tons\ Wood}{Year}\right) * (75\%Yield) * \left(\frac{\$3}{Gal}\right) = \mathbf{\$675,000/Yr}$ **Selling Terpenes**	Sell Terpenes **$675,000**
6) Reduce Maintenance Costs:[28] Media changeouts, large heavy moving part replacement, corrosion repairs, etc.	Save up to **~$1 Million**
Total Value Creation per Year per Site:	**Up to $3 Million**

---

22 ICAC Guidance Method for Estimation of Gas Consumption in a RTO. Institute of Clean Air Companies website. https://cdn.ymaws.com/www.icac.com/resource/resmgr/RTO-F1.pdf (Accessed September 18, 2021)

23 "Operating Cost Calculator". Cycle Therm website. http://www.cycletherm.com/resources/operating-cost-calculator.aspx (Accessed November 21, 2021).

24 "Greenhouse Gases Equivalencies Calculator - Calculations and References". Environmental Protection Agency (EPA) website. https://www.epa.gov/energy/greenhouse-gases-equivalencies-calculator-calculations-and-references (Accessed December 28, 2020).

25 "Allowance Price Explorer". International Carbon Action Partnership website. https://icapcarbonaction.com/en/ets-prices (Accessed Jan 20, 2021).

26 "Price Chart for Gum Turpentine". PineChem website. http://www.pinechem.net/images/gall/ChartGNE.GIF Accessed Jan 3, 2021).

27 "Sulfate Turpentine Recovery". Pulp Chemicals Association. Pages 9-12, 127. https://www.pinechemicals.org/store/viewproduct.aspx?id=1579188 (Accessed Jan 2, 2021).

28 "Air Pollution Control Technology Fact Sheet". EPA website. https://www3.epa.gov/ttncatc1/dir1/fregen.pdf (Accessed October 17, 2021).

VIII. CASE STUDY: FBC TECHNOLOGY SPECIFIC TO WOOD DRYING

DEMONSTRATION DESCRIPTION:

Captis Aire LLC founder and Environmental C&C Inc. worked on a project together from 2015-2021 with a commercial oriented strand board (OSB) manufacturer to demonstrate the ability to utilize the FBC to control emissions from wood drying.

In wood products manufacturing, process gas exhaust streams released from drying wood contain VOCs, including turpentine (terpenes). These VOCs are considered pollutants/emissions and must be removed from the air, prior to release, to avoid air pollution emissions into our communities. Wood drying operations utilized in the manufacture of oriented strand board (OSB), lumber, wood to biofuels, i.e. wood pellets, plywood, particleboard, etc. are few examples of these processes that have process exhaust streams containing VOCs.

Work on this patent[29] pending[30] technology was executed in three phases. In the first phase, laboratory screening work was done to explore whether emissions from wood drying, primarily terpenes, could be adsorbed by Bead Activated Carbon (BAC) media at the expected operating temperatures. In the second phase, a small-scale pilot adsorber was installed at a commercial Oriented Strand Board (OSB) manufacturing site to demonstrate ability to adsorb emissions from wood drying under commercial manufacturing conditions. In the third phase, the pilot unit capabilities were expanded to include the adsorber, desorber, and condenser. The SSR performance was simulated by semi-continuous additions of virgin BAC during air emissions compliance testing.

In the third testing phase, the pilot unit ran for 6 months with over 1700 hours on-line. Ability to meet air emissions control compliance, per the Plywood and Composite Wood Products Maximum Achievable Control Technology (PCWP MACT) standard, was determined by two independent certified air emissions[31] stack testing companies[32] during two separate testing events.

---

[29] "METHODS AND SYSTEMS FOR RECOVERING TERPENE COMPOSITIONS FROM WOOD DRYING EXHAUST". World International Patent Office (WIPO) Website. https://patentscope.wipo.int/search/en/detail.jsf?docId=WO2019148208 (Accessed September 18, 2021).

[30] "METHODS AND SYSTEMS FOR CONTROLLING EMISSIONS FROM WOOD DRYING PROCESSES". WIPO Website. https://patentscope.wipo.int/search/en/detail.jsf?docId=WO2019148207 (Accessed September 18, 2021).

[31] "AIR Advanced Industrial Resources". AIR website. https://airtest1.com/ (Accessed April 16, 2020).

[32] "Testing and Modeling". John Zink Hamworthy Combustion website. https://www.johnzinkhamworthy.com/testing-modeling/ (Accessed Feb 3, 2021).

TABLE V
RESULTS OF AIR POLLUTION CONTROL COMPLIANCE TESTING (FBC UNIT):

Emissions Measured:	FBC %Reduction Efficiency[33]
Total Hydrocarbons (THC), as carbon including Methane	92%
Volatile Organic Compounds (VOCs), as carbon. (Total non-methane hydrocarbons – THC minus Methane)	92%
Formaldehyde	83%
Methanol	75%
GHC - Nitrogen Oxides (NOx)	55%
Particulate Matter (PM), total filterable	50%
Condensable Particulate Matter	59%
Total Particulate Matter, Filterable + Condensable	51%
Total PM2.5, Particle Size Analyses (PSA) Filterable + Condensable	60%
Total PM10, Particle Size Analyses (PSA) Filterable + Condensable	64%
Carbon Monoxide (CO)	7%
Sulfur Dioxide (SO2)	99%

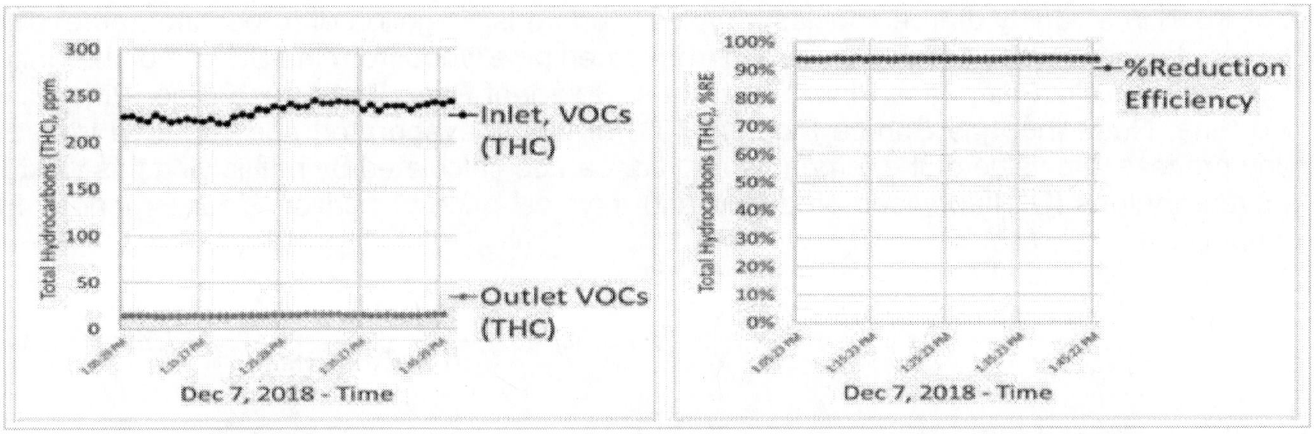

RESULTS OF VOC REDUCTION EFFICIENCY PERFORMANCE MEASURED BY FLAME IONIZATION DETECTOR (FBC DEMONSTRATION UNIT):
Achieved >90% Reduction Efficiency meeting PCWP MACT Standard[34]

---

33 Emissions results FBC 2018 at commercial wood drying site done by Advanced Industrial Resources, Inc. Results were demonstrated during three 1-hour testing intervals. https://airtest1.com/ . (Accessed September 18, 2021)

34 "Electronic Code of Federal Regulations (eCFR)". eCFR website. See https://www.ecfr.gov/cgi-bin/text-idx?node=sp40.13.63.dddd#ap40.14.63_12292.1 (Accessed April 16, 2020).

RESULTS OF VOLATILE ORGANIC COMPOUNDS CAPTURE:

Terpenes were collected from the FBC demonstration unit. These terpenes are a mixture of terpene hydrocarbons, primarily Alpha and Beta pinenes. These terpenes are responsibly-recovered, biobased materials. Potential applications for these terpenes include use in fuels, fragrances, solvents, adjuvants, cleaners, and polyterpene resins.

TERPENES COLLECTED FROM FBC DEMONSTRATION UNIT

VALUE OF TERPENES:

The Pine Chemicals Association has documented the gallons of terpenes per oven dry ton of wood in a variety of tree species. Typically, there is ~1 gallon of turpentine (terpenes)[35] per oven dry ton of wood product produced from green pine wood from the SE part of the United States, i.e. Loblolly Pine, Jack Pine, Slash Pine, Shortleaf Pine, Ponderosa Pine, Pitch Pine, Pond Pine, Red Pine, and Spruce Pine. If 75% of this is evaporated and captured from the drying process, the value of the terpenes per year can be calculated by multiplying this yield by the oven dry tons (ODT) of wood produced and a typical price per gallon[36] for terpenes. For example:

$$\left(\frac{1\ Gallon\ Terpenes}{ODT\ Wood}\right) * \left(\frac{ODT\ Wood}{Year}\right) * (\%Yield) * \left(\frac{\$\ Dollars}{Gallon\ Terpenes}\right) = \$\ /\ Year$$

TABLE VI
TERPENES TYPICAL PROPERTIES:

Property	Result
Appearance	Clear Liquid
Moisture	<0.1%
Weight per Gallon	7.2 pounds per gallon
Specific Gravity	0.86
Sulfur Content	< 10 ppm

---

35 "Sulfate Turpentine Recovery". Pulp Chemicals Association. Pages 9-12, 127. https://www.pinechemicals.org/store/viewproduct.aspx?id=1579188 (Accessed Jan 2, 2021).
36 "Price Chart for Gum Turpentine". PineChem website. http://www.pinechem.net/images/gall/ChartGNE.GIF Accessed Jan 3, 2021).

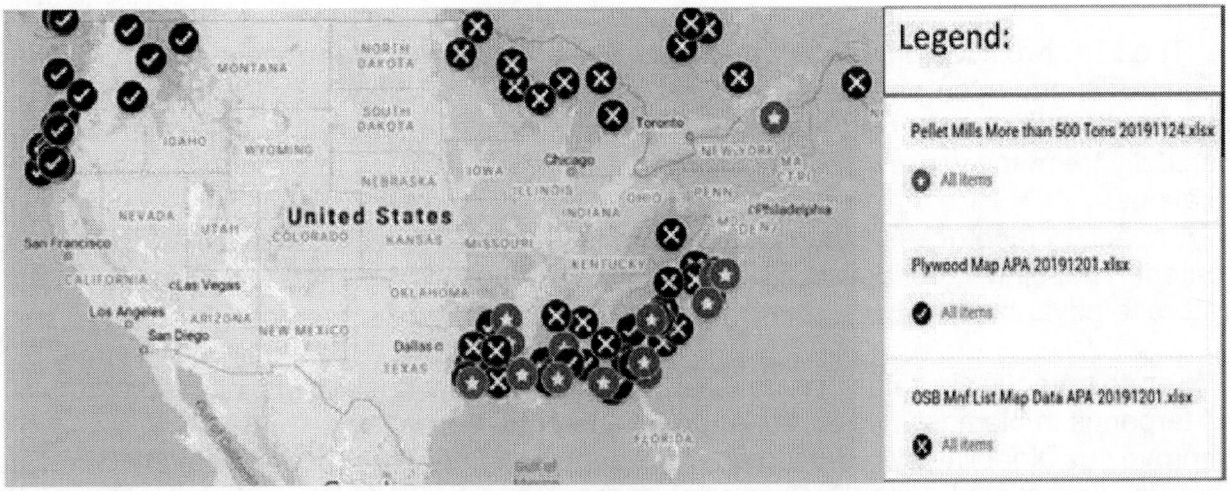

MAP OF POTENTIAL WOOD DRYING SITES FOR FBC USE:
This map shows potential sites where various wood products are produced where the FBC may be used to control emissions. Potential sites include but are not limited to Oriented Strand Board (OSB), lumber (sawmills)[37], plywood[38], wood to biofuels, i.e. wood pellets[39], particleboard, Medium Density Fiberboard (MDF), etc.

RESULTS OF DISCUSSIONS WITH ENVIRONMENTAL AGENCIES:

Official stack testing results were discussed with two state environmental agencies and two federal environmental protection agency (EPA) personnel. In all discussions, environmental agency personnel were supportive of providing a permit for full size commercial FBC unit.

FINAL RESULTS FROM WOOD DRYING DEMONSTRATION. THE UNIT:

- Provided salable terpenes, composed primarily of Alpha and Beta Pinenes.
- Reduced greenhouse gas (GHG) emissions.
- Reduced energy usage. It eliminated natural gas usage and reduced electrical usage.
- Met the PCWP MACT requirement of >90% Reduction Efficiency[40] in both tests. In the second test, the system met the compliance requirements three times in a row, over each of the three one-hour testing intervals.
- Demonstrated good operational reliability.
- Enabled State and Federal environmental personnel that reviewed these results to support permitting for a full commercial FBC system.

---

37 "The mill map allows you to visually locate and drill down to find the mill you're interested in." Primary Forest Products Network website. https://primary.forestproductslocator.org/mill-map (Accessed September 18, 2021).

38 "Manufacturer Directory". Engineered Wood Association website. https://www.apawood.org/manufacturer-directory (Accessed September 18, 2021).

39 "Biomass Magazine". Biomass Magazine website. See http://biomassmagazine.com/plants/map/pellet/ (Accessed September 18, 2021).

40 "Electronic Code of Federal Regulations (eCFR)". eCFR website. See https://www.ecfr.gov/cgi-bin/text-idx?node=sp40.13.63.dddd#ap40.14.63_12292.1 (Accessed April 16, 2020).

IX. CONCLUSIONS:

The FBC provides an affordable, reliable, and accessible clean technology for controlling VOCS, HAPS, and even particulate in some applications. This technology is well established in many applications including automotive and electronics manufacturing where it has been used successfully for many years. It is an emerging technology for use in biomass (i.e. wood drying) applications.

Especially for process exhaust streams with low VOC concentrations, the FBC technology is a patent[41] pending[42], cost competitive, industrial air pollution control technology that enables operators to advantageously:

1. Generate revenue via sales of recovered organic products (including biobased products such as terpenes in biomass applications) and sales of carbon offsets
2. Improve sustainability by reducing energy usage and thus greenhouse gas (GHG) emissions
3. Reduce operational costs by reducing energy usage (natural gas and electricity) and maintenance costs.
4. Enhance VOC mass transfer to the BAC media
5. Provide continuous in-situ BAC media regeneration and reactivation. This enables the media to continue to be reused.
6. Minimize water adsorbed on the BAC media thus maintaining its ability to adsorb VOCs, even in high moisture content process exhaust streams

Like any air pollution control technology installation, careful consideration of many design parameters, the unit's specific operating profile, and the operator's goals need to be taken into account during the design, installation, and startup/optimization processes. There are excellent resources, including Environmental C&C Inc.[43], Captis Aire LLC[44], and the Institute of Clean Air Companies (ICAC)[45] personnel, to assist with evaluating, implementing, and optimizing FBC applications.

---

[41] "METHODS AND SYSTEMS FOR RECOVERING TERPENE COMPOSITIONS FROM WOOD DRYING EXHAUST". World International Patent Office (WIPO) Website. https://patentscope.wipo.int/search/en/detail.jsf?docId=WO2019148208 (Accessed September 18, 2021).
[42] "METHODS AND SYSTEMS FOR CONTROLLING EMISSIONS FROM WOOD DRYING PROCESSES". WIPO Website. https://patentscope.wipo.int/search/en/detail.jsf?docId=WO2019148207 (Accessed September 18, 2021).
[43] "Environmental C&C Inc". Environmental C&C website. https://www.environmentalcc.com/ (Accessed September 18, 2021).
[44] "Emissions Control Technology Captures Emissions to Clean the Air". Captis Aire website. https://www.captisaire.com/ (Accessed November 21, 2021).
[45] "The Trusted Voice of the Clean Air Technology Industry". Institute of Clean Air Companies (ICAC) website. https://www.icac.com/ (Accessed September 18, 2021)

X. ACKNOWLEDGEMENTS:

The author recognizes inputs from a broad spectrum of personnel at the commercial OSB manufacturer and at Environmental C&C[46] including Harold Cowles, John Berger P.E., and Jim Starek.

CAPTIS AIRE CAPTURE-CONCENTRATE-CLEAN	Environmental C&C, Inc. Solvent Recovery & VOC Abatement Systems
Kim TUTIN Captis Aire LLC [47] 1612 W. Cleveland Ave, Suite 001 East Point, GA 30344-3213 Tel: +01 404 5802795, E-mail: kim@captisaire.com Website: www.captisaire.com	Harold COWLES Environmental C&C Inc. 16705 E. Calle de las Estrellas Road Scottsdale, AZ 85262-5982 Tel: +01 518 2250315, E-mail: halc@ecnc.com Website: https://www.environmentalcc.com/

[46] "Environmental C&C Inc". Environmental C&C website. https://www.environmentalcc.com/ (Accessed September 18, 2021).
[47] "Emissions Control Technology Captures Emissions to Clean the Air". Captis Aire website. https://www.captisaire.com/ (Accessed November 21, 2021).

631638 Conversion of Plastic Waste to Fuel: Assessment of Technologies and Economics

Abstract:

Plastics are extremely valuable to our society and are widely used across industries such as food, beverage, textile, healthcare, automotive, aerospace, construction, agriculture, and oil. The global production of plastics, accumulative from 1950 to 2017, is 8.3 billion metric tons, with 6.3 billion tons waste generated, of which only 9% has been recycled, 12% incinerated, with the remaining 79% landfilled or released into the environment. Mechanical recycling is used to recycle about 95% of plastic waste today, however, this method works well only for homogeneous and non-contaminated plastic waste, and for easily identifiable objects such as bottles and containers made of PET and PE. Chemical recycling can offer a solution in recycling many of the plastics which mechanical recycling fails to process. Chemical recycling involves the deconstruction of polymers to monomers for conversion back into polymers, or to intermediates that can be used as fuel or feedstock by the chemical industry. A number of chemical recycling technologies are being developed, and the processes are being tailored to handle specific feedstocks and/or produce desirable products. In this project we assess companies that are involved in the chemical recycling of plastic waste: their technology and/or key innovations, the feedstocks and product(s), and the current scale of operation. Further, we evaluate in this project the economics of chemical recycling with a specific interest on plastic-to-fuel conversion.

623136 Continuous Back to Monomer Recycling of PET from Composite Materials Using a Twin-Screw Extruder

Abstract:

The current Packaging Act requires a 63% quota for the mechanical recycling of plastics from 2022 in Germany [1]. In order to meet this requirement, new efficient recycling technologies are necessary. Many plastics, such as polyethylene terephthalate (PET), are used in combination with other polymers. Apart from energy recovery, no recycling process for such multilayer material is available and it is therefore largely lost to the resource cycle.

The research project revolPET® specifically addresses this issue through the development of a continuous process for recovering the monomers terephthalic acid (TA) and ethylene glycol (EG) in primary raw material quality from mixed PET-containing waste [2]. While literature mainly reports batchwise depolymerization of PET in stirred tank reactors [3, 4], the revolPET® process uses an extruder as a continuous chemical reactor.

This contribution presents experimental results on the successful depolymerization of PET in composite materials into a TA salt and EG. The high flexibility of the extruder offers a wide range of options for process design in terms of thermal, mechanical and chemical conditions, so that suitable process windows are identified depending on the feed material. Key to this technology, the twin-screw extruder allows the two reagents to be mixed as solids. This innovative process intensification enables high degrees of depolymerization at average residence times below 2 minutes. In the experiments, a PET depolymerization degree of $\geq$ 97 % was achieved. Systematic parameter studies reveal the influence of temperature, filling degree of the screw chamber and residence time on the successful PET depolymerization. Figure 1 presents typical results for degree of depolymerization and residence time at varied screw revolutions.

In addition to the technical aspects, the contribution provides insight into the ecological assessment of this new recycling process. In particular, the challenge of the exothermic reaction to identify an optimal process window is shown. This topic is examined in more detail against the background of scaling. The presented experimental results will be discussed in detail and placed in the context of the current political debate.

Literature

[1] § 16 Absatz 2 VerpackG

[2] www.revolpet.eu

[3] A.M. Al-Sabagh, F.Z. Yehia, Gh. Eshaq, A.M. Rabie, A.E. ElMetwally, *Egyptian Journal of Petroleum* (2016) 25, 53-64, doi: 10.1016/j.ejpe.2015.03.001

[4] A. Oku, L.-C. Hu, E. Yamada, *J. Appl. Polym. Sci.* (1997) *63*, 595-601.

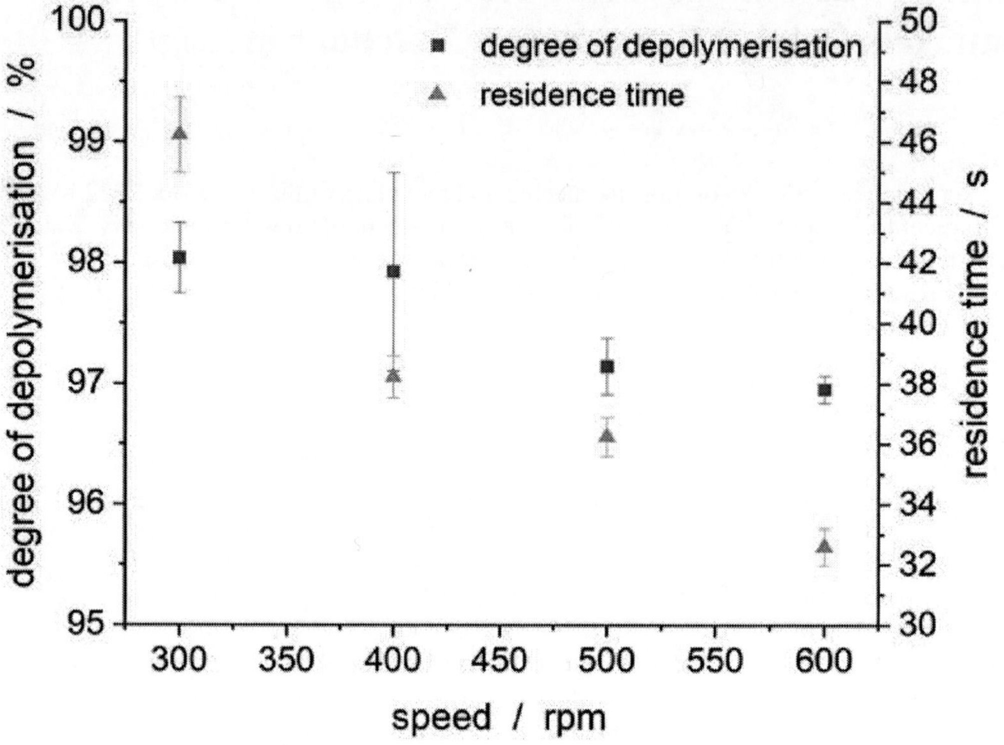

Figure 1: Degree of PET depolymerization in relation to the residence time at different speed

629642 The Effect of Residence Time and Heterogeneous Catalysis on Continuous Hydrothermal Liquefaction of Food Waste

Abstract:

The United States releases a significant amount of methane through the decomposition of waste in landfills; methane is a greenhouse gas with a global warming potential 25 times greater than carbon dioxide [1]. According to the EPA, landfilling accounts for approximately 17% of the total US methane emissions [1], and food waste is the largest non-recyclable component of municipal waste [2]. Although food waste currently harms the environment, it also presents an opportunity to solve another looming threat: the US dependency on non-renewable fossil fuels. Mainstream Engineering, in collaboration with Worcester Polytechnic Institute, is developing technology to convert food waste to energy through hydrothermal liquefaction (HTL) that simultaneously addresses our reliance on fossil fuels and the environmental impact of food waste.

HTL converts food waste, and other wet wastes, into an energy dense biocrude by reacting the food mixture of carbohydrates, lipids, and proteins in near supercritical water. Traditionally, HTL reactions have been uncatalyzed or used a homogeneous catalyst (i.e., Na_2CO_3). We have shown previously that using a solid oxide, heterogeneous catalyst, $CeZrO_x$, in a batch reactor improves biocrude yield and energy recovery [3]. We have subsequently studied the effects HTL residence time and catalyst addition in a continuous stirred tank reactor (CSTR). A model food waste slurry was fed to the CSTR via a dual high pressure syringe pump system. The pressure of the system was maintained using a back pressure regulator rated to 300 °C and 5,000 PSI. Commercially available $CeZrO_x$ powder was compressed into tablets and immobilized in the CSTR using a catalyst cage. The catalyst tablets were able to withstand the hydrothermal environment without physically disintegrating. The stability of the catalyst tablets was characterized using XRD and BET before and after HTL. The energy recovery, product yields, and biocrude properties (HHV, TAN, and CHNO) were determined to characterize the effectiveness of $CeZrO_x$ in a continuous HTL reactor and evaluate the impact of residence time on continuous HTL of food waste.

References

[1] EPA. *Greenhouse Gas Emissions*. [cited 2021; Available from: https://www.epa.gov/ghgemissions/overview-greenhouse-gases.

[2] USDA. *Why should we care about food waste?* [cited 2021; Available from: https://www.usda.gov/foodlossandwaste/why.

[3] Maag, A.R., et al., *Catalytic hydrothermal liquefaction of food waste using CeZrOx*. Energies, 2018. **11**(3): p. 564.

629415 Continuous Hydrotreatment of Sewage Sludge and *Spirulina* Algae Biocrudes from Hydrothermal Liquefaction: Different Catalysts for Different Organic Contaminants

Abstract:

Unprecedented global warming and raising public awareness are the driving factors to decarbonize long-haul transportation sector. In this regard, biomass could be a potential renewable source for the production of sustainable, carbon-neutral liquid fuels. Hydrothermal liquefaction (HTL) followed by catalytic hydrotreatment is one of the most promising pathways to covert varying feedstocks into drop-in fuels. The energetically dense black viscous product from HTL ("biocrude"), contains a diverse organic pool with considerable amount of oxygen (~5-18%), nitrogen (~1-8%) and metallic content (~0.1-0.6%). The presence of these organic contaminant results in poor miscibility with fossil fuels and poses new and exciting challenges e.g. thermal instability of biocrude at high temperatures which are indispensable for complete hydrodenitrogenation. This is why special attention is needed not only to realize this complex biocrude mixture but also to select different catalysts which performs best to saturate and remove certain organic contaminants under optimized conditions.

In this work, we utilized advanced analytical tools (i.e. FT-ICR MS etc.) and observed different classes of organometallics and nitrogen compounds from sewage sludge and *Spirulina* algae biocrude. From these results, we utilized three different types hydrotreating catalysts based on their porosity and nickel loading on γ-Al_2O_3 support and successfully demonstrate smooth continuous hydroprocessing operations (Figure 1) for hundreds of hours without coke formation, catalyst deactivation, and reactor plugging. Finally, 100% deoxygenation and ~96% denitrogenation were achieved. Furthermore, hydrotreated products underwent true-boiling point fractional distillation and showed on-spec metal content, density, pour point, cloud point and HHV for diesel and jet fuel.

.

Figure 1: Hydrotreated samples from the continuous hydrotreatment of sewage sludge (a) and *Spirulina* algae (b) biocrude.

Acknowledgement: This research has received funding from the European Union's Horizon 2020 research and innovation programme under grant agreement No 764734.

627420 Review and Modeling of Organic Removal in Anaerobic Sludge Blanket Bioreactors for Methane Production and Energy Recovery from Domestic Wastewater

Abstract:

The current domestic wastewater treatment paradigm centers on aerobic wastewater treatment technologies such as activated sludge. While effective in meeting U.S. EPA regulated discharge standards for organics and suspended solids, providing oxygen to aerobic microorganisms is energy-intensive, typically accounting for ~50% of wastewater treatment facility costs. In total, wastewater conveyance and treatment accounts for ~3% of our nation's electrical energy usage. Direct anaerobic treatment of dilute wastewater is a viable methane-generating alternative to aerobic wastewater treatment. Anaerobic microorganisms convert wastewater organics (fats, carbohydrates, proteins, etc.) to methane-rich biogas, which can be harnessed to create energy via technologies such as combined heat & power. If more energy is generated from the methane produced than is used to run the process itself, then the process can be considered "energy-positive" (or a net energy producer). Anaerobic sludge blanket bioreactors, most notably upflow anaerobic sludge blanket (UASB) bioreactors and anaerobic baffled reactors (ABRs), have been studied at bench- and pilot-scale for several decades. UASBs are also currently operated at full-scale in several locations worldwide, mostly in warmer climates. Despite the number of completed studies, no current review critically examines performance data from different sized sludge blanket bioreactors collected under varying conditions (e.g., varying wastewater temperatures and hydraulic residence times). Further, no current review study conducts predictive modeling (i.e., Monte Carlo Simulation) on aggregated sludge blanket bioreactor data to better understand the operation of such bioreactors under varying conditions. This study aggregates and examines data from 145 UASBs studies and 56 ABRs studies to understand bioreactor performance (e.g., chemical oxygen demand and total suspended solids removal) and methane generation. This study also employs Monte Carlo Simulation using @Risk to better predict bioreactor performance under varying conditions. Additionally, this study identifies critical gaps in literature (e.g., a lack of consistently reported methane generation) that may impact full-scale implementation in certain climates (e.g., cold weather climates). Last, results from this study can help inform decision makers to better understand of anaerobic sludge blanket bioreactors can meet regulatory standards or generate methane for renewable energy at their wastewater treatment facilities.

629993 Utilization of Marble Dust As a Concrete Additive to Enhance Its Properties

Abstract:

With the advancement in the building technology, each and every step to alter the cement properties for making it environmentally beneficial and equally strong is important. In this advancement, one always opts for an alternative which provides the strength to the building by reducing its cost and weight.

Marble often known as calcium carbonate ($CaCO_3$) has been a well known material used as a building material since the ancient times. Marble is often available in combination with Silicon Oxide (SiO_2), Aluminum Oxide (Al_2O_3), Iron Oxide (Fe_2O_3), Magnesium Oxide (MgO) and Calcium Oxide (CaO).

Marble is utilized at a number of operations and the waste obtained requires an extra effort for the adequate environmental disposal. The waste marble left in the environment can be harmful therefore must be utilized for an alternative process to reduce its harmful effect. The addition of marble dust to concrete is proving to be an effective alternative for the process. The process of marble addition is participating in the hydraulic reactions contributing significantly to the composition and microstructure of hydrated products. Another way of utilizing the marble dust is to utilize it in the geo-polymer concrete which will be strengthening the concrete properties.

The entire paper is evaluating the utilization of the marble dust in concrete so that the unsafe waste of marble is used and fewer efforts is to be taken for its disposal and also to accomplish viability, economy and environmental conservation.

622829 Oxygen Non-Stoichiometry and Defect Models of Brownmillerite-Structured $Ca_2MnAlO_{5+\delta}$ for Chemical Looping Air Separation

Abstract:

The brownmillerite-structured $Ca_2MnAlO_{5+\delta}$ has demonstrated excellent oxygen storage capacity that can be used for chemical looping air separation (CLAS), a potentially efficient approach to produce high-purity oxygen from air. In order to effectively utilize this material as an oxygen sorbent in CLAS, it is necessary to comprehensively understand its thermodynamic properties and the structure-performance relationships in the operating range of interest. In this work, the oxygen non-stoichiometry (δ) of $Ca_2MnAlO_{5+\delta}$ was systematically measured by thermal gravimetric analysis (TGA) in the temperature ranging from 440-600°C and under oxygen partial pressure 0.01-0.8 atm. The partial molar enthalpy and entropy for oxygen releasing reaction were calculated using van't Hoff Equation with an averaged value of 146.5 ± 4.7 kJ/mol O_2 and 162.7 ± 5.1 J/K mol O_2, respectively. The experimentally measured non-stoichiometry (δ) was well-fitted by a point defect model applied in two regions divided by the predicted equilibrium *P-T* curve. The equilibrium constants for appropriate defects reactions were also determined. The thermochemical parameters, molar enthalpy and entropy for the main reaction, obtained from the defect model were 141.7 kJ/mol O_2 and 139.1 J/K mol O_2 respectively, showing agreement with the aforementioned values. It thus demonstrated the constancy on theoretical calculations as well as the high fidelity of the selected defect model. The resulting defect model was also applied to higher oxygen partial pressure environment in the range from 1-4 atm and exhibited satisfactory fitting quality. The experimental studies on oxygen non-stoichiometry combined with the defect-structure modeling provide useful insights on the fundamental understanding of the oxygen sorbents' redox performances and helpful information for the design and optimization of oxygen sorbents in CLAS.

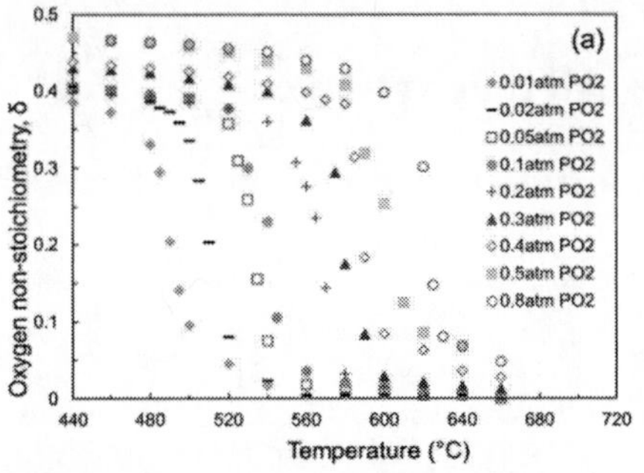

Figure (a). Oxygen non-stoichiometry (δ) of $Ca_2MnAlO_{5+\delta}$ at different temperatures and P_{O_2}

Figure (b). Experimentally measured oxygen non-stoichiometry (δ) of $Ca_2MnAlO_{5+\delta}$ (square points) and defect model predictions (dashed lines) as a function of P_{O_2} at fixed temperatures in low P_{O_2} region and **(c)** in high P_{O_2} region.

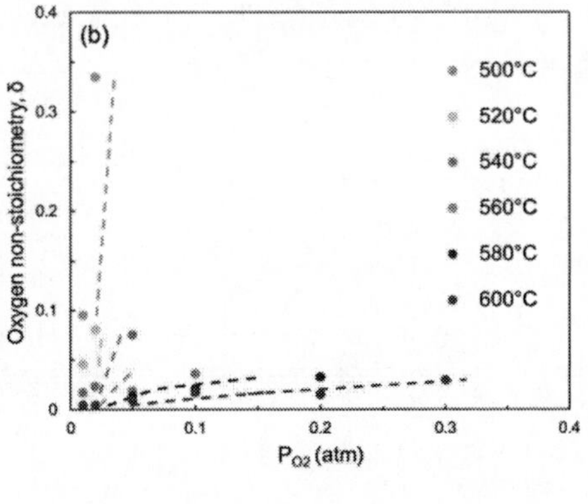

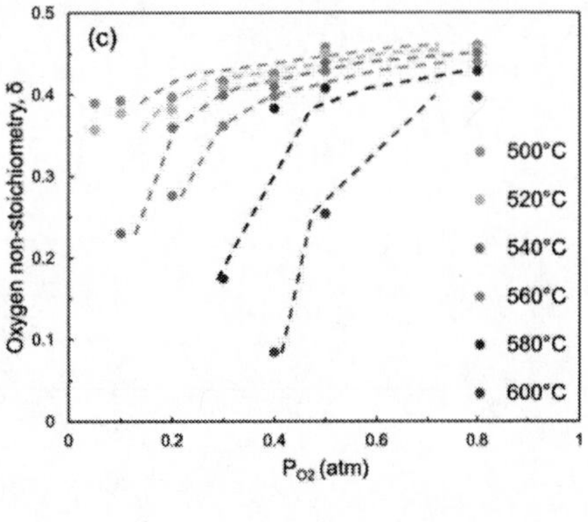

623357 Life Cycle Assessment of a Chemical Looping Hydrogen Production Process with Varied Reducer, Oxidizer, and Combustor Pressures

Abstract:

H_2 is an essential industrial chemical and a promising clean fuel for the future. However, the current methods for H_2 production, such as steam methane reforming (SMR), can lead to significant amount of CO_2 emissions, and the global warming potential (GWP) of a typical SMR process is approximately ~12 kg CO_2eq/kg H_2. In order to reduce the CO_2 emissions associated with H_2 production, CO_2 capture units can be integrated into SMR processes. However, this usually results in considerable reduction in energy efficiency and H_2 yield of the process. This research studies a chemical looping process that converts natural gas into H_2 with high energy efficiency and 100% CO_2 capture. The process is simulated and optimized in ASPEN Plus to maximize H_2 yield while achieving autothermal operation. Then, the life cycle assessment (LCA) is performed on this process to compare its global warming potentials with the conventional SMR processes with and without CO_2 capture. The calculation shows that the chemical looping process only has a GWP of ~5 kg CO_2eq/kg H_2. This GWP can be further reduced when a varied pressure operating strategy is implemented into the chemical looping process, allowing the reducer, the oxidizer, and the combustor to operate at different pressures. With this strategy, the GWP can be further reduced to 3 kg CO_2eq/kg H_2. This research demonstrates the great advantage of chemical looping technology in reducing the greenhouse gas emissions associated with H_2 production from traditional fossil fuels.

624710 Techno-Economic Evaluation of H_2 Production Using Chemical Looping Dry Reforming

Abstract:

Chemical looping dry reforming (CLDR) is a novel process for production of inherently separated syngas (H_2 and CO) streams. In CLDR, CH_4 is first converted to a pure hydrogen stream and solid carbon via thermocatalytic cracking over a metal catalyst in a "cracker" reactor. The deposited carbon is then selectively removed in an oxidizer reactor using CO_2 as oxidant, resulting in the production of high-purity CO and regenerating the metal which is recycled back to the cracker reactor. The metal thus acts as a "carbon carrier" as opposed to an "oxygen carrier" as in typical chemical looping processes. Overall, CLDR results in the formation of inherently separated syngas streams, while at the same time addressing some of the key shortcomings of conventional methane cracking, such as low CH_4 conversion, rapid catalyst deactivation, and large temperature excursions during carbon burn-off in the oxidation half cycle.

In the present study, we perform a techno-economic evaluation of a CLDR process that is designed to maximize H_2 production by combining CLDR with water gas shift (WGS) in order to utilize the produced CO for further H_2 production. Different configurations are evaluated for supplying the heat required for the endothermic dry reforming reaction via combustion of additional CH_4 or combustion of a fraction of the carbon produced in the cracker reactor using either pure oxygen or air. All configurations include CO_2 capture and H_2 purification via pressure swing adsorption. Building on our previous fixed-bed experimental results, reactor-level mass and energy balance calculations are performed and coupled with a systems-level model that incorporates thermodynamic calculations of individual process components such as water gas shift reactors, blowers, feed compressors, CO_2 capture, PSA, and heat recovery equipment as well as steam turbines. Capital cost models for the major equipment and operating costs (catalysts, chemicals, fuel etc.) are also developed using data in the open literature. The results from the performance model are combined with the cost models to comparatively evaluate the three configurations based on performance and cost metrics such as energy efficiency, chemical efficiency, CO_2 emissions, capital costs, and cost of H_2 production. The results are compared with conventional H_2 production pathways using steam reforming and dry reforming of methane. The results are used to identify areas where CLDR systems have technical and cost advantages over conventional processes and highlight factors which need improvement.

631031 Chemical Looping CO_2 splitting and Methane Reforming for Low-Carbon-Footprint Chemical Synthesis

Abstract:

Conversion of CO_2 to CO for chemical production is one of the most promising routs for widescale carbon utilization. However, traditional CO_2 to CO technologies pose challenges. Dry reforming of methane (DRM) which co-produces hydrogen is one of the most widely studied but has several drawbacks, including high energy demand, high temperatures (>800 °C), and coke formation. The 1:1 H_2:CO syngas product of dry reforming is useful for several chemicals, but few commercial processes can directly convert a 1:1 syngas without costly syngas separation, and/or water-gas shift (WGS). Due to the challenges of DRM, industrial syngas producers typically use methane partial-oxidation/autothermal reforming, followed by energy intensive separation to obtain high concentration CO feedstocks for synthesis (which offers no CO_2 utilization benefit). We present chemical looping catalyst for a hybrid redox process (HRP) that produces separate CO and syngas streams. The syngas is produced by partial oxidation of methane by the lattice oxygen of the HRP catalyst. In a subsequent step the lattice oxygen is regenerated by CO_2 splitting. Through choosing mixed metal oxides with catalytically active surfaces and/or modifying the particle surfaces with catalytically active metals, HRP offers high conversions in both reaction steps. Compared to dry reforming, HRP permits the efficient utilization of heat directly produces concentrated CO and 2:1 H_2:CO streams without parasitic separations/WGS reactions.

We present experimental results that show over 95 % CO yield in the CO_2-splitting step and >95% selectivity towards syngas with a near 2:1 H2:CO in the partial oxidation step. The activity and selectivity of the redox catalysts during both the reduction and oxidation steps were tested. The performance of both platinum group metal (PGM)-containing and PGM-free redox catalyst are presented. PGM-free catalyst are shown to maintain excellent activity and stability in HRP operation, greatly enhancing the economic feasibility of the approach. Stable performance is observed over 500 cycles/125 hours with the phases intact. Simulations based on experimental HRP results are briefly presented indicating large cost and net CO_2 savings for production. HRP is an ideal choice for sustainable CO_2 fixation due to the wide industrial use for CO in chemical production.

AUTHOR INDEX

AIChE
120 Wall Street, FL 23
New York, NY 10005-4020

ISBN 978-1-7138-5686-3
9 781713 856863

CURRAN ASSOCIATES INC.
proceedings.com